Ronny Vavrin

Time-resolved Investigation of Aggregation and Gelation of Colloids

Ronny Vavrin

Time-resolved Investigation of Aggregation and Gelation of Colloids

Time-resolved Investigation of Aggregation and Sol-Gel Transition in Colloidal Suspensions

Südwestdeutscher Verlag für Hochschulschriften

Impressum/Imprint (nur für Deutschland/ only for Germany)

Bibliografische Information der Deutschen Nationalbibliothek: Die Deutsche Nationalbibliothek verzeichnet diese Publikation in der Deutschen Nationalbibliografie; detaillierte bibliografische Daten sind im Internet über http://dnb.d-nb.de abrufbar.

Alle in diesem Buch genannten Marken und Produktnamen unterliegen warenzeichen-, marken- oder patentrechtlichem Schutz bzw. sind Warenzeichen oder eingetragene Warenzeichen der jeweiligen Inhaber. Die Wiedergabe von Marken, Produktnamen, Gebrauchsnamen, Handelsnamen, Warenbezeichnungen u.s.w. in diesem Werk berechtigt auch ohne besondere Kennzeichnung nicht zu der Annahme, dass solche Namen im Sinne der Warenzeichen- und Markenschutzgesetzgebung als frei zu betrachten wären und daher von jedermann benutzt werden dürften.

Verlag: Südwestdeutscher Verlag für Hochschulschriften Aktiengesellschaft & Co. KG
Dudweiler Landstr. 99, 66123 Saarbrücken, Deutschland
Telefon +49 681 37 20 271-1, Telefax +49 681 37 20 271-0, Email: info@svh-verlag.de
Zugl.: Fribourg, Uni, Diss., 2005

Herstellung in Deutschland:
Schaltungsdienst Lange o.H.G., Zehrensdorfer Str. 11, D-12277 Berlin
Books on Demand GmbH, Gutenbergring 53, D-22848 Norderstedt
Reha GmbH, Dudweiler Landstr. 99, D- 66123 Saarbrücken
ISBN: 978-3-8381-0514-7

Imprint (only for USA, GB)

Bibliographic information published by the Deutsche Nationalbibliothek: The Deutsche Nationalbibliothek lists this publication in the Deutsche Nationalbibliografie; detailed bibliographic data are available in the Internet at http://dnb.d-nb.de.

Any brand names and product names mentioned in this book are subject to trademark, brand or patent protection and are trademarks or registered trademarks of their respective holders. The use of brand names, product names, common names, trade names, product descriptions etc. even without
a particular marking in this works is in no way to be construed to mean that such names may be regarded as unrestricted in respect of trademark and brand protection legislation and could thus be used by anyone.

Publisher:
Südwestdeutscher Verlag für Hochschulschriften Aktiengesellschaft & Co. KG
Dudweiler Landstr. 99, 66123 Saarbrücken, Germany
Phone +49 681 37 20 271-1, Fax +49 681 37 20 271-0, Email: info@svh-verlag.de

Copyright © 2008 Südwestdeutscher Verlag für Hochschulschriften Aktiengesellschaft & Co. KG and licensors
All rights reserved. Saarbrücken 2008

Produced in USA and UK by:
Lightning Source Inc., 1246 Heil Quaker Blvd., La Vergne, TN 37086, USA
Lightning Source UK Ltd., Chapter House, Pitfield, Kiln Farm, Milton Keynes, MK11 3LW, GB
BookSurge, 7290 B. Investment Drive, North Charleston, SC 29418, USA
ISBN: 978-3-8381-0514-7

Contents

Abstract   5

I   Simultaneous time-resolved Neutron/DWS/DLS   7

1 Theory   11
   1.1   Colloidal suspensions and gels 11
       1.1.1   Colloidal suspensions 12
       1.1.2   Aggregation and gelation 14
   1.2   Methods to characterise gels 19
       1.2.1   General description of scattering experiments 20
       1.2.2   Light scattering 30
       1.2.3   Neutron scattering 49

2 System   51
   2.1   Choice of System 51
   2.2   Method of destabilisation (urea/urease) 54
   2.3   Sample preparation for the investigation of time resolved aggregation and gelation 55

3 Setup   57
   3.1   DWS 57
       3.1.1   Non ergodic DWS 58
   3.2   SANS 62
   3.3   combined setup 63

       3.3.1   Experimental settings used in the aggregation and gelation experiments 64

4 Aggregation, cluster formation and sol-gel transition in a moderately concentrated colloidal suspension   67
   4.1   Structure S(q) 68
   4.2   Dynamics 70

5 Discussion   77
   5.1   Highly correlated initial suspension 77
   5.2   S(q) scaling method 84
   5.3   DWS analysis 87
       5.3.1   DWS applied on small particles 88
       5.3.2   Calculation of l^* from SANS data for DWS 90
       5.3.3   DWS applicability at low L/l^* 93
       5.3.4   Mean square displacement $\langle \Delta r^2 \rangle$ 95
   5.4   Gelation kinetics 96
       5.4.1   Additional Echo DWS measurements 101

6 Finite gels with varying Φ 109

7 Conclusion   115

II   Multi3D   119

8 Setup   123
   8.1   Theory 3D DLS and SLS 123
       8.1.1   Intercept β and overlap volume 126
       8.1.2   SLS on turbid media 127
   8.2   Description setup 128
   8.3   Software 132
   8.4   Alignment 133
       8.4.1   Mechanical alignment 133

	8.4.2 Optical alignment	136
9	**Test measurement**	**139**
10	**Further development/outlook**	**141**
Acknowledgments		**151**

Abstract

In this work we present a study of the aggregation and sol-gel transition of colloidal suspensions using a wide range of experimental techniques, where some of them were modified or developed especially for this purpose. Covering all relevant time and length scales, we investigate in the first part of this work a model system which undergoes a sol-gel transition in the hope to gain further insight into the fundamental process of gelation. Therefore, a colloidal suspension of nano particles (polystyrene spheres) in water is destabilised to induce aggregation of fractal clusters. The clusters will grow until they finally all connect and form a volume filling network, i.e. a gel. Combining neutron and light scattering, we are able to measure simultaneously and time resolved the static and dynamic properties of the same sample in the length scale of a few Angstroms to several hundred nanometers, which allows us to test well-known theories and to compare our data with recent results of other research groups. Our measurements demonstrate a deviation from classical theories in the critical behavior just before the volume spanning network is formed, which we interpret with the observation of a glassy cluster phase. In order to investigate this effect, a novel technique is used to gain access to very slow dynamic processes, revealing data which seems to proof our hypothesis of a glassy phase in the gelation process.

The second part of this work describes an experimental setup which is based on a light scattering technique introduced in the last ten years and which allows us to measure faster and with higher precision. By measuring simultaneously at four different angles instead of only one, a factor of four in acquisition time is gained in comparison to traditional setups. Moreover, dynamic measurements can be analysed angle dependent and therefore improved in precision. As the chosen technique is based on a multiple scattering suppression scheme, even relatively turbid samples can be analysed using the well established single scattering theory. Therefore the developed setup is an ideal tool to investigate the sol-gel transition under conditions hardly accessible otherwise.

Part I

Simultaneous time-resolved Neutron scattering/DWS/DLS: gelation of colloidal suspensions

Our aim in this part is to study time resolved gelation processes of colloidal suspensions. This is done by a combination of two non invasive complementary techniques: a specially designed setup includes light and neutron scattering for simultaneous measurements of dynamical and structural properties of destabilized nanoparticle suspensions and gels. We will describe the temporal evolution of the structure and dynamic properties of destabilized particle suspensions over a large range of length and time scales. We monitor the initial cluster growth, the crossover from diffusive motion to network fluctuations at the gel point and the subsequent evolution of the network properties with time.

Chapter 1

Theory

1.1 Colloidal suspensions and gels

Gels are very well known in every day life applications and they play an important role in industrial and biological processes. Soft cheese and yoghurt are two examples of products which underlie the same mechanisms like the production of ceramics via modern sol-gel processing or blood coagulation.

A gel is best described by a network of particles suspended in a liquid medium. This network spans the whole sample volume and, due to its attractive inter-particle forces, results in an elastic, solid-like behavior of the sample, even when small forces are applied. Depending on the aggregation mechanism, gel formation can occur even at very low volume fractions, leading to a soft solid with measurable elasticity.

A particle gel is formed, when colloidal particles suspended in a liquid medium connect to each other and build a network. Frequently, gel formation occurs via the formation of open fractal clusters due to irreversible aggregation of the individual particles. The clusters grow and eventually become space filling at the so-called gel point.

In the following two subsections I will briefly review the most important theoretical concepts used to describe the properties of colloidal suspensions and gels.

1.1.1 Colloidal suspensions

A colloidal suspension consists of a small solid particles suspended in a liquid medium, further on referred to as solvent. The solid particles are called colloids and are of the size of between 1nm and 1μm. Lyophilic colloids like proteins are thermodynamically stable suspensions whereas lyophobic colloids like organic or inorganic particles such as latex (polystyrene, PS) or Titanium dioxide (TiO_2) are thermodynamically unstable. In the lyophobic case, kinetically "trapped" suspensions can only exist due to a stabilisation mechanism. Typical examples are charge-stabilised particles or colloids with an attached polymer layer that provides a so-called steric stabilisation. A classical example of colloidal suspensions are small gold particles suspended in water which Michael Faraday prepared in 1857 and which are still stable.

Sub micron sized particles in a colloidal suspension do not sediment or fall out because of their small size. They are small enough to undergo Brownian motion, and the thermal energy overcomes gravitational forces. As they move in a random way, particle collisions are inevitable. In the presence of attractive interparticle interactions such as due to van der Waals or hydrophobic forces, particles would stick together and aggregate to bigger clusters, and the suspension would be unstable and the particles would all sediment out with time. For this reason, all particles in a stable colloidal suspension have to be stabilised by repulsive interactions. There are two main mechanisms to avoid aggregation: steric and charge stabilisation. In the case of steric stabilisation, polymers are adsorbed at the surface of the particles. These polymers build a hairy layer around the particle and repel other hairy particles because a penetration of the polymer layers constrains the number of possible polymer configurations. This leads to a rise of the entropy and therefore to a repulsion of the particles. In our work, we have mainly focussed on the second mechanism, charge stabilisation. For charged colloids, the interparticle potential is often modeled with a so-called DLVO (Derjaguin-Landau-Verwey-Overbeek) potential

$$V(s) = V_C(s) + V_{VdW} \qquad (1.1)$$

where the repulsive screened Coulomb potential $V_C(s)$ is combined with the attractive long-range van der Waals potential V_{VdW}. s stands for the surface to surface distance between the particles and a for the particle radius. As ionisable groups contained in the particle dissociate in polar solvent, a diffuse layer of counterions is formed around the particle. In the case of $s \ll a$, the resulting Coulomb potential can be simplified to

$$V_C(s) = 2\pi\epsilon a \zeta^2 \cdot \ln(1 + e^{-\kappa \cdot s}) \quad (1.2)$$

with ζ being the Zeta-potential of the particles (which is the interaction potential of the particles observed by hydrodynamic measurements), ϵ the dielectric constant of the solvent and κ the inverse of the screening length. κ is given by $\kappa = \sqrt{4\pi e^2/(\epsilon k_B T) \cdot \rho |Z|}$ where ρ is the number density of the particles and $|Z| = Q/e$ is the valency of the particles. By adding further electrolytes with a number density ρ_I and valency $|Z_I|$ to the suspension, κ can be calculated by

$$\kappa = \sqrt{\frac{4\pi e^2}{\epsilon k_B T}(\rho|Z| + 2\rho_I|Z_I|)} \quad (1.3)$$

For larger distances s, the screened Coulomb potential can be calculated in a more general approach which leads to the so-called Yukawa potential $V_Y(s)$

$$V_Y(s) = \frac{Q^2}{\epsilon\left(1 + \frac{\kappa a}{2}\right)^2} \frac{\exp^{-\kappa(r-a)}}{r} \sim \frac{1}{r}\exp^{-\kappa r}; \quad r > a \quad (1.4)$$

where r stands for the distance between the centers of the particles. The Van der Waals potential for spherical particles is given by

$$V_{VdW} = -\frac{A}{6} \cdot \left(\frac{2a^2}{s^2 + 4as} + \frac{2a^2}{s^2 + 4as + 4a^2} + \ln\left(\frac{s^2 + 4as}{s^2 + 4as + 4a^2}\right)\right) \quad (1.5)$$

where A is the Hamaker constant. The Hamaker constant A describes the strength of the attractive Van der Waals potential and depends on the dielectric functions of the particles and the solvent.

Due to the DLVO potential, particles with high surface charge densities or weakly screened charges feature a potential barrier which keeps them separated, illustrated in figure 1.1. A secondary minimum can also be present, but it becomes only significant if particles of different radii interact (particularly when $a_1/a_2 > 10$) or if the radius of the particles of a monodisperse suspension is large ($a > 50nm$ for particles with surface charge densities of $1.8\mu C/cm^2$ in a suspension containing 50 mM ions). A detailed discussion of the lower inset follows in the next chapter.

1.1.2 Aggregation and gelation

In order to form a gel, the particles of a colloidal suspension have to aggregate i.e. to stick together and eventually connect to a network spanning the whole sample volume. Therefore the stabilisation mechanism mentioned before has to be overcome. In the case of charge stabilisation, the Coulomb potential can easily be modified through the addition of salt. By adding ions to the suspension, the particle charges are screened (particle charges are neutralised by a double layer of counter-ions). On account of this, the Coulomb potential weakens and becomes more and more short ranged; the resulting DLVO potential is shown in the lower inset of figure 1.1. The high interaction energy barrier of strongly repelling particles a) will decrease in height and shorten in range (b), c)). Below a certain barrier height, the thermal energy $k_B T$ starts to be sufficiently high for a certain percentage of the particles to overcome the barrier, and particles start to aggregate at a slow rate. Once the screening is strong enough to suppress the energy barrier completely, the so-called "critical coagulation concentration" (ccc) is reached (d)). From this point on, the particles aggregate immediately upon contact, resulting in a fast aggregation rate.

The quantity to describe the stability of a colloidal system is the stability ratio W which is defined as $W = k_r/k$ where k_r and k are the rate constants

---

[1]This figure was published in Intermolecular and Surface Forces With Applications to Colloidal and Biological Systems, 2nd Edition, Jacob Israelachvili, Page 248, Copyright Academic Press (1991).

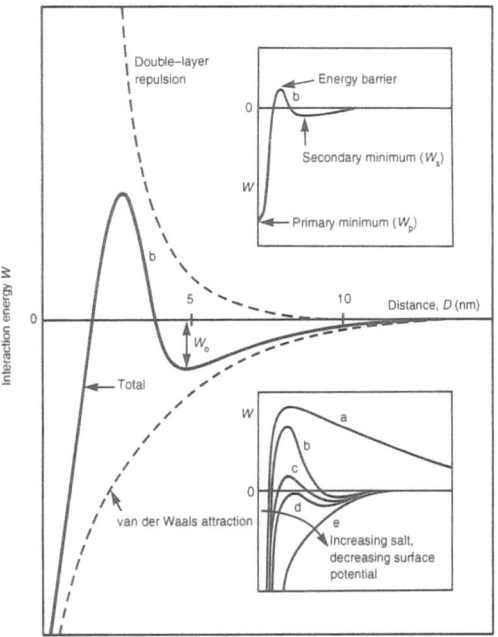

Figure 1.1: DLVO as described by Israelachvili[1]. The total potential (interaction energy W) is the sum of the Coulombx repulsion (double layer repulsion) and the Van der Waals attraction. a) particle surfaces repel strongly; the colloidal suspension is stabilised. b) and c): the energy barrier gets lower and more short-ranged due to screening, colloids coagulate slowly (RLCA). d) the "critical coagulation concentration" (ccc) is reached; the energy barrier vanished completely, colloids start to coagulate rapidly (DLCA). e) no charge repulsion anymore, rapid aggregation.

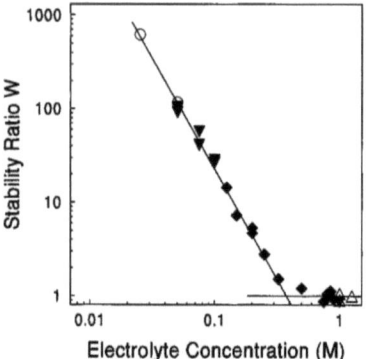

Figure 1.2: Stability ratio W of polystyrene latex spheres.[2] The slope of the slow aggregation at low electrolyte concentration M changes at the critical coagulation point ccc to the plateau of the fast aggregation regime.

for the rapid resp. measured aggregation. Figure 1.2 shows the stability ratio W against salt concentration [1]. At low salt concentrations, the aggregation is slow and the aggregation rate depends on the salt concentration. The dependance between logW and log electrolyte concentration is close to linear in this regime. At the critical coagulation concentration ccc, the curve turns parallel to the concentration axis. From this point on, fast aggregation takes place. The fast aggregation does not depend on the salt concentration anymore; the charges of the particles are totally screened and the particles will connect irreversibly immediately upon contact.

There is a model with two basic mechanisms describing the formation and growth of clusters: Diffusion Limited Cluster Aggregation (DLCA) and Reaction Limited Cluster Aggregation (RLCA). For DLCA the particles will create a bond immediately upon each contact (sticking probability = 1),

---

[2]Reprinted with permission from H. Holthoff, S. U. Egelhaaf, M. Borkovec, P. Schurtenberger, H. Sticher, Langmuir 12(23), 5541-5549 (1996). Copyright (1996) American Chemical Society.

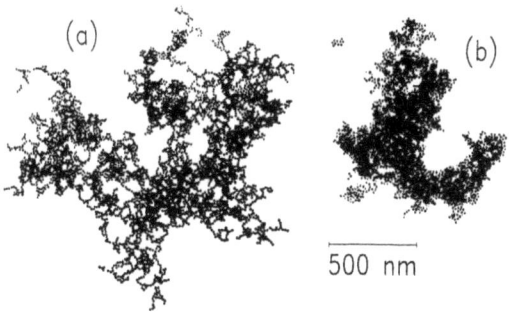

Figure 1.3: TEM pictures of aqueous gold colloids[3] with a very uniform particle radius of a=7.5nm. a) fractal DLCA structure b) more compact, fractal RLCA structure

forming very loose fractal clusters. This mechanism sets in after the salt concentration in the suspension reached the ccc point. In the case of RLCA, the salt concentration is lower than the ccc and several contacts are needed until the particles stick together (sticking probability $\ll 1$). The particles therefore can penetrate further into a cluster, and as a consequence the RLCA structures are more dense and compact (see figure 1.3).

Both DLCA and RLCA cluster structures are irregular, but can successfully be interpreted using fractal geometry [2, 3]. Fractals [4] are self-similar structures with a non-integer fractal dimension d_f that links the cluster radius R_c and cluster mass M via $M \propto R_c^{d_f}$. The theoretically predicted fractal dimension for DLCA is $d_f \approx 1.8$ and for RLCA $d_f \approx 2.1$. The cluster radius shows for DLCA a power-law growth with time t

$$R_c \propto t^{\frac{1}{d_f}} \qquad (1.6)$$

---

[3]Reprinted figure with permission from D. A. Weitz, J. S. Huang, M. Y. Lin, and J. Sung, Phys. Rev. Lett. 54, 1416 - 1419 (1985). Copyright (1985) by the American Physical Society.

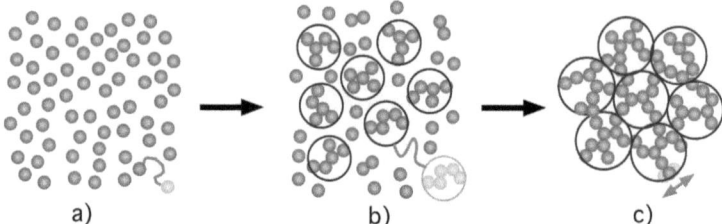

Figure 1.4: Stages of a gelation process: a) stable suspension with particles in brownian motion; b) destabilised suspension with aggregating particles and growing clusters, all still undergoing brownian motion; c) one big space filling cluster with particles bound to the network, only able to wiggle around its average position in the network.

whereas the cluster grow in an exponential way in the RLCA regime

$$R_c \propto e^{\alpha t} \tag{1.7}$$

The sequence of particle destabilisation and subsequent cluster formation and finally sol-gel transition is shown schematically in figure 1.4. The stable suspension of charged colloids is illustrated in a). The particles are undergoing Brownian motion. By screening the particle charges, the potential barrier vanishes and the destabilised suspension starts to build aggregates. The aggregates steadily grow but are still small enough do diffuse freely in the suspension (part b)). When the clusters finally grow and interconnect to one space filling cluster as in c), a network is spanned through the whole sample. The particles then are trapped in the network and cannot explore the whole phase space but can only wiggle around an average position in the network. Such a system then shows elastic behavior. The addition of salt to a charged colloidal suspension is a simple way to induce aggregation and gelation. Some technical problems of this simple approach will be discussed in section (2.2).

The critical cluster radius $R_{c,critical}$ for gel formation, i.e. the point where the suspension of clusters becomes space filling and a gel is formed, can be estimated with

$$R_{c,critical} \approx a\Phi^{-1/(3-d_f)} \tag{1.8}$$

where a is the radius of one single colloidal particle and Φ the volume fraction. Usually, an experimentally estimated pre-factor is added to this rough estimate. As an example, Weitz et al. found in the case of $1 \cdot 10^{-4} \leqslant \Phi \leqslant 5 \cdot 10^{-3}$ and for $d_f \approx 2.1$ a pre-factor of $a = 0.3$ for charged polystyrene spheres with a radius of 9.5nm [5]. It is important to realise that the fractal regime in particle gels has lower and upper limits, here the particle and the critical radius of the clusters at the point when they interconnect, $R_{c,critical}$. Particle gels will then exhibit fractal properties only in the regime $a \ll \xi \ll R_{c,critical}$, whereas at length scales $\xi > R_{c,critical}$ they are homogeneous and non-fractal (ξ stands for the correlation length).

1.2 Methods to characterise gels

Scattering experiments are widely used to characterise materials and are perfectly suited for soft condensed matter [6]. When an incident beam of light, x-rays or neutrons hits a sample, it can either pass trough the sample, be absorbed or be scattered. By measuring the scattered intensities it is possible to obtain information on different properties; on the one hand, the time averaged intensity as a function of the scattering angle (so-called static scattering) is a fingerprint of the structure and osmotic compressibility of the sample. When the length of the scattering vector $\vec{q}$ (definition see equation (1.11)) is extrapolated to zero, the mass of a particle resp. the osmotic compressibility of the sample can be obtained from static scattering. The radius of gyration R_g, the form factor $P(q)$ and the structure factor $S(q)$ (explained in the following subsections) can be extracted from the q-dependance of the static scattering experiment, where $P(q)$ describes the shape of an individual particle and its internal mass distribution and $S(q)$ the spatial correlations between the particles. On the other hand, the time dependent fluctuations of the scattered intensity can be evaluated (so-called dynamic scattering) in order to learn about the dynamics of the scatterers i.e. the motion of the particles. Here, we will focus on light and neutron scattering, as these apply best for our chosen system.

1.2.1 General description of scattering experiments

The following concepts are based on several assumptions: First of all, the scattering process is supposed to be quasi-elastic, neglecting any absorption or inelastic scattering with significant energy transfer. Additionally, the incident beam should not be distorted significantly by the medium in order to satisfy the Born approximation of the first order. Furthermore, the incident beam is regarded as a plane wave and the scattered beam as a spherical wave. Moreover, the scattering centers are small compared to the wavelength of the scattered beam. And finally, the distance between detector and scattering center is meant to be sufficiently large (detection in the far field).

Single point scatterer

Under the assumptions mentioned above, the amplitude of the spherical wave scattered by one scattering center at rest can be described as

$$A_s(\vec{R'}) = A_0 b \frac{e^{i\vec{k}_s \vec{R'}}}{|\vec{R'}|} \tag{1.9}$$

where R' is the distance between the scattering center and the detector and $\vec{k}_s$ the scattered wave vector (see left side of figure 1.5). The scattered amplitude A_s is then composed of the amplitude of the incident beam A_0, the so-called scattering length b and a term representing a spherical wave. The scattering length b describes the interaction between the incident beam and the scattering center.

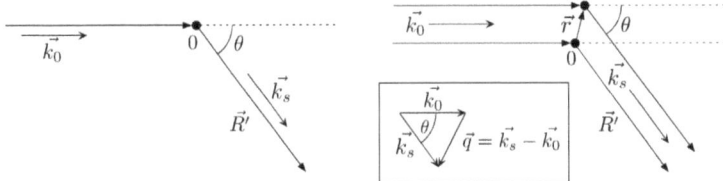

Figure 1.5: Left: single point scatterer, right: two point scatterers. The distance between scattering center and detector is R'. Inset: definition of the scattering vector $\vec{q}$.

Two and more point scatterers

When an incident beam hits two scattering particles at rest (see right side of figure 1.5), interference effects will appear. The scattered amplitude is then given by

$$A_s(\vec{R'}) = \sum_{j=1}^{2} A_s^j \simeq \frac{A_0}{R'} e^{i\vec{k}_s \cdot \vec{R'}} \sum_{j=1}^{2} b_j e^{i\Delta\varphi_j} \qquad (1.10)$$

The right part of the equation is only valid when the distance R' is much bigger than the distance r between both scattering centers (this is the so-called far field approximation). If one point scatterer is positioned in the center of a coordinate system, one phase shift factor zeroes out ($\Delta\varphi_1 = 0$), and only $\Delta\varphi_2$ remains. The latter then depends only on the path length Δs between both scattering centers. By introducing the so-called scattering vector

$$\vec{q} = \vec{k}_s - \vec{k}_0 \qquad (1.11)$$

which is the difference between the scattered and incident wave vectors $\vec{k}_s$ and $\vec{k}_0$ (see inset in figure 1.5), $\Delta\varphi_2$ can be rewritten as

$$\Delta\varphi_2 = \frac{2\pi}{\lambda}\Delta s = \vec{q} \cdot \vec{r} \qquad (1.12)$$

For elastic scattering ($|\vec{k}_s| = |\vec{k}_0|$), the scattering vector can be expressed as

$$q = |\vec{q}| = \frac{4\pi \sin(\frac{\theta}{2})}{\lambda} \qquad (1.13)$$

where θ designates the scattering angle and λ the wavelength of the incident beam for this medium. Using this nomenclature, equation (1.10) then looks as follows

$$A_s(\vec{R'}) = \frac{A_0}{R'} e^{i\vec{k_s}\cdot\vec{R'}} \sum_{j=1}^{2} b_j e^{i\vec{q}\vec{r_j}} \qquad (1.14)$$

and is easily generalised from two to N particles:

$$A_s(\vec{R'}) = \frac{A_0}{R'} e^{i\vec{k_s}\cdot\vec{R'}} \sum_{j=1}^{N} b_j e^{i\vec{q}\vec{r_j}} \qquad (1.15)$$

Scattering intensity, differential cross section and scattering length

Detectors in experimental setups are not sensitive to scattered amplitudes but measure the scattered intensity. The latter can be expressed by

$$\left\langle I_s(\vec{R'}) \right\rangle_e = \left\langle A_s(\vec{R'}) \cdot A_s^*(\vec{R'}) \right\rangle_e = \frac{A_0^2}{R'^2} \sum_{j,k=1}^{N} \left\langle b_j b_k e^{i\vec{q}(\vec{r_j}-\vec{r_k})} \right\rangle_e \qquad (1.16)$$

where $\langle\rangle_e$ stands for the average over the full ensemble. To normalise for detector distance R' and the incident intensity I_0, the so-called differential scattering cross section $\frac{d\sigma}{d\Omega}$ is defined by

$$\frac{d\sigma}{d\Omega}(\vec{q}) = \frac{\left\langle I_s(\vec{R'}) \right\rangle_e}{I_0} R'^2 = b^2 \sum_{j,k=1}^{N} \left\langle e^{i\vec{q}(\vec{r_j}-\vec{r_k})} \right\rangle_e \qquad (1.17)$$

assuming in the second step that all particles are equal ($b_j = b_k = b$). In the case of colloids, the background scattering of the solvent (with intensity $I_{s,solv}$) has to be substracted, and the equation above is slightly modified to

$$\frac{d\sigma}{d\Omega}(\vec{q}) = \frac{\left\langle I_s(\vec{R'}) \right\rangle_e - \left\langle I_{s,solv}(\vec{R'}) \right\rangle_e}{I_0} R'^2 \approx \Delta b^2 \sum_{j,k=1}^{N} \left\langle e^{i\vec{q}(\vec{r_j}-\vec{r_k})} \right\rangle_e \qquad (1.18)$$

where $\Delta b = b - b_{solv}$ designates the so-called excess scattering length, with b_{solv} as the scattering length of a molecule of the solvent. Such a simple subtraction of the background is strictly valid only in the case of incompressible suspensions.

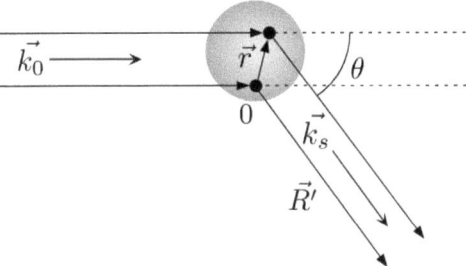

Figure 1.6: Schematic representation of two point scatterers which are part of a single particle (grey sphere).

The scattering length b of a point scatterer is related to the differential scattering cross section by

$$b^2 = \frac{d\sigma}{d\Omega}(\vec{q}) \qquad (1.19)$$

and can have a positive or negative value. A negative value signifies a phase change of the scattered wave (the scattered wave is shifted half a wave length in comparison to the incident wave). As the scattering length defines the interaction between scattering center and the incident wave, it varies for different probing beams (here light and neutrons) and will be treated in the corresponding subsections.

Usually, the scattering lengths b of all nuclei b_j in a volume V_0 are summed to the average scattering length density

$$\rho = \frac{1}{V_0} \sum_j b_j \qquad (1.20)$$

respectively the background-corrected excess scattering length density

$$\Delta\rho = \frac{1}{V_0} \sum_j \Delta b_j \qquad (1.21)$$

Scattering from a single particle at rest

A particle can be regarded as an arrangement of many point scatterers (see figure 1.6). The scattered intensity is the sum of the spherical waves emitted by all the point scatterers, and as these waves can interfere, characteristic scattering patterns can be observed for particles of different sizes and shapes. In the case of a dilute non-interacting suspension, the correlation between the individual single particles can be neglected. Assuming N identical particles, the total scattered intensity can then simply be expressed as the sum of the intensities $I_p(q)$ of the single particles

$$I_S(q) = N \cdot I_p(q) \tag{1.22}$$

This can also be written in the following form

$$I_S(q) = N \cdot I_p(0) P(q) \quad \text{with} \quad P(q) = \frac{I_p(q)}{I_p(0)} \tag{1.23}$$

where the form factor P(q) is introduced. The form factor tends by definition to 1 when $\vec{q} \to 0$ and describes the q-dependence of the scattering from one particle due to intraparticle interference effects. In the case of a homogeneous sphere, the RGD form factor (in the Rayleigh-Gans-Debye limit) is easily caculated:

$$P(q) = \left(\frac{3(\sin qa - qa \cos qa)}{(qa)^3} \right)^2 \tag{1.24}$$

where the form factor P(q) of a homogeneous sphere has minima around qa=4.49, 7.73, ... (see figure 1.7). The RGD regime is a special case of very weakly scattering and/or small particles where the condition $2k_0 a |n_p/n_s - 1| \ll 1$ (where a is the particle radius) is fulfilled. While for neutron scattering the RGD limit is always observed, for light scattering the mismatch between the index of refraction of the particle n_p and of the solvent n_s can be substantial, and we have to consider other scattering regimes. Light scattering beyond this limit is described by Mie theory [7, 8]. For very big particles (above several micrometers), the electromagnetic wave cannot penetrate into

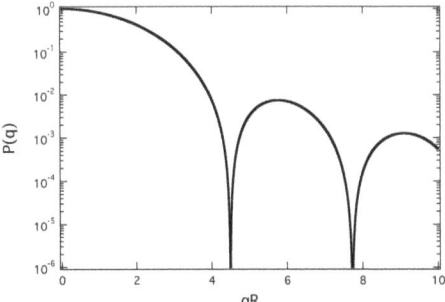

Figure 1.7: RGD form factor in a log-lin plot against qR; the minima at 4.49 and 7.73 are well visible.

the particle and the scattering process can be approximated by the interaction of a planar wave with the cross section of the particle. This limiting case is known as Fraunhofer scattering.

The expansion of the form factor given in the equation above to a power series in the limit of small values of qa results in the so-called Guinier approximation

$$P(q) \cong 1 - \frac{1}{3}q^2 R_G^2 + O(q^4) \qquad (1.25)$$

Using this approximation, the radius of gyration R_G of any object (particles of any form and structure) can be measured at low q ($(qR_G)^2 \ll 1$). R_G is the radius of gyration of the object, defined by the root-mean square distance of the mass elements from their centre of mass.

Scattering from many particles at rest

Considering a concentrated system or particles with long-range interactions, we cannot longer neglect the inter-particle interaction. Assuming the N particles with centres of mass at position $\vec{R}_j(t)$ to be identical homogeneous spheres, we can write the total scattered intensity (using equation (1.23)) as

$$I_S(q) = N \cdot I_p(0)P(q) \sum_{j,k=1}^{N} \left\langle e^{-i\vec{q}\cdot(\vec{R}_j - \vec{R}_k)} \right\rangle = N \cdot I_p(q)P(q)S(q) \qquad (1.26)$$

introducing the structure factor $S(q)$

$$S(q) = \frac{1}{N} \sum_{j,k=1}^{N} \left\langle e^{-i\vec{q}\cdot(\vec{R}_j - \vec{R}_k)} \right\rangle \qquad (1.27)$$

The scattered intensity of the N individual particles without correlation between them is already given in equation (1.23). Therefore, all correlation effects caused by the spatial arrangement of the particles are incorporated in the structure factor $S(q)$. In the case of a dilute system where the interactions are negligible, the structure factor becomes 1. The particle pair correlation function $g(r)$ (radial distribution function) describes the arrangement of particles in real space. It is related to the structure factor through Fourier inversion

$$g(r) = 1 + \frac{1}{2\pi^2}\left(\frac{N}{V}\right) \int_0^\infty (S(q) - 1)q^2 \frac{\sin(qr)}{qr} dq \qquad (1.28)$$

Three examples of different structures with corresponding $S(q)$ and $g(r)$ are shown in figure 1.8.

In our case, two structures need to be considered. In the initial charge stabilised colloidal suspension, the long range and weakly screened Coulomb repulsion leads to very strong positional correlations between the particles and a structure factor that closely resembles that from a super-cooled liquid. This structure factor features a peak at q^* which is linked to the average distance $\langle d \rangle$ between particles by $q^* \approx 2\pi/\langle d \rangle$ and scales with $\phi^{-1/3}$ (as the interparticle distance can be calculated from the particle volume fraction ϕ using $\langle d \rangle = \sqrt[3]{V_p/\phi}$ where V_p is the particle volume). Upon increasing the salt concentration, the degree of correlation will decrease, the structure

---

[4]Reprinted figure with permission from C. Urban, PhD thesis, Swiss Federal Institute of Technology Zurich, 1999 (ISBN-13: 978-3-89675-622-0). Copyright (1999) by Herbert Utz Verlag GmbH.

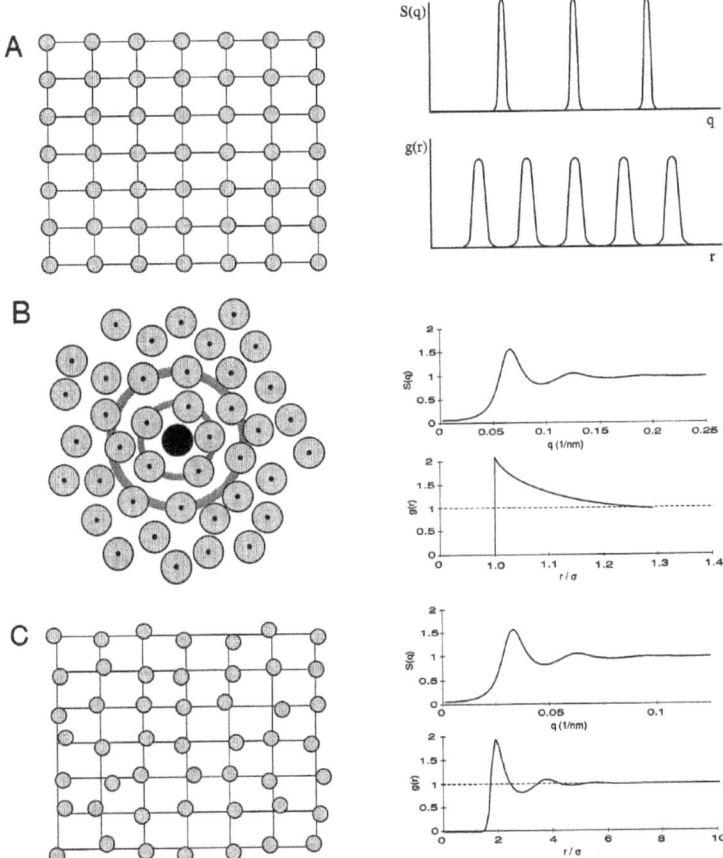

Figure 1.8: Structure factors $S(q)$ and pair correlation functions $g(r)$ of three different particle arrangements: A) crystal like, B) hard sphere suspension and C) charge stabilised suspension with long range interactions.[4]

factor peak disappear and the forward scattering increase. As soon as the particles start to aggregate, this leads to enhanced forward scattering and to a qualitative change in $S(q)$ due to the existence of large clusters with fractal structure. As mentioned in (1.1.2), the characteristic feature of a fractal structure is its self-similarity. This is reflected in a power law dependence of the scattered intensity with the scattering vector

$$I \propto q^{-d_f} \qquad (1.29)$$

In the final stage when the gel is formed by the space filling network, the structure factor shows both the fractal structure of the network caused by the aggregated particles but also the finite size of the critical cluster size $R_{c,critical}$ when the cluster connect to a space filling network. The finite cluster size is reflected in the Guinier approximation (eqation (1.25)) with which the size of any scattering object can be measured at low q. At q close to or bigger as $1/R_G$, the Guinier approximation is not applicable anymore due to the fact that the measurement then resolves the internal, fractal structure of the cluster. The Fisher-Burford structure factor combines fractal structure and Guinier approxomation to

$$S(q) = \left(1 + \left(\frac{2}{3d_f}\right) q^2 R_{c,critical}^2\right)^{\frac{-d_f}{2}} \qquad (1.30)$$

The equation above takes into account that the gel network consists of a homogeneous closed packed array of fractal blobs i.e. is homogeneous on length scales $1/q > R_{c,critical}$, but fractal on the length scale $1/q < R_{c,critical}$. In figure 1.9, the structure factors $S(q)$ during aggregation and gelation of a DLCA particle gel with volume fraction $\Phi = 0.02$ are shown (calculated using Monte Carlo Simulations by Rottereau et al. [9]). With increasing gel time t_g, the fractal clusters grow and scatter with rising intensity in the small q range. The growing size and fractal structure of the clusters is reflected in the increasing length of the slope which corresponds to the fractal dimension $d_f = 1.8$ (straight dashed line) at lower q values. At high q values, the structure factor shows a damped oscillation which is a consequence of the

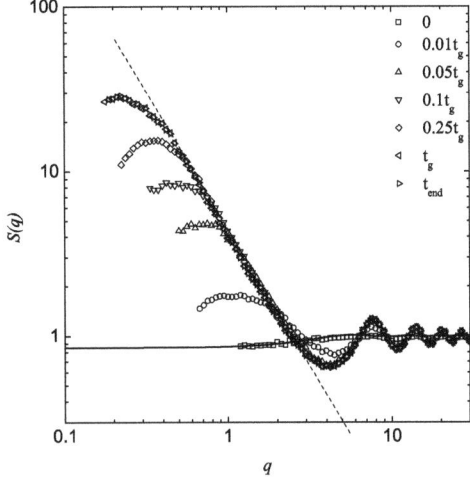

Figure 1.9: Time resolved structure factors $S(q)$ for a DLCA particle gel (Φ=0.02), calculated using Monte Carlo Simulations[5]. t_g is the "gel time" where the whole sample volume is filled with one cluster. The dashed line has a slope of -1.8, the expected value for the fractal dimension of a DLCA gel. The solid line represents the theoretical result for hard spheres.

structure of the first and second shell around a particle. The structure factor of the gel therefore sinks below the value of the structure factor of random hard spheres (solid line) at certain q values behind the fractal slope.

As a general remark, it is important to mention that the scattered intensity measured at a scattering vector q is maximal when the phase difference $\vec{q} \cdot \vec{r}$ between scatterers is a multiple of 2π (due to constructive interference similar to Bragg peaks). For a given value of q, the scattered intensity is therefore governed by the intensity scattered from scatterers which are sepa-

---

[5]Reprinted figure with permission from M. Rottereau, J. C. Gimel, T. Nicolai, and D. Durand, European Physical Journal E, 15:141148, 2004 (http://epje.edpsciences.org/). Copyright (2004) by EDP Sciences.

rated by a distance of $r = 2\pi/q$. As a rule of thumb, scattering experiments executed at the scattering vector q probe the sample with a spatial resolution of $2\pi/q$. At small angles and therefore small q, the probed length scale is large and whole particles or even macroscopic structures can be measured. Experiments at high q reflect the arrangement of scatterers at short length scales and are able to resolve the inner structure of particles. Special emphasis should be placed on the fact that the measured intensity in all scattering experiments represents the Fourier transform of the spatial correlation function of the scattering length densities i.e. we investigate the sample in reciprocal space.

1.2.2 Light scattering

The light scattering techniques which are traditionally most used are based on the principle of single light scattering (static and dynamic light scattering, explained in the next two paragraphs). The single light scattering condition is fulfilled when a system scatters so weakly that the probability of a photon being scattered twice or more is negligible. This is ensured by diluting the sample or by using systems with small scattering contrast. Furthermore, we shall consider mostly conditions where the incident beam is not distorted significantly by the medium i.e. when the Raleigh-Gans-Debye (RGD) condition $2k_0 a |n_p/n_s - 1| \ll 1$ is fulfilled (a is the particle radius, n_p and n_s the refractive indices of particle and solvent and k_0 the wave number $k_0 = 2\pi/\lambda$). Under these circumstances, static and dynamic scattering can be described by relatively simple and quantitative expressions.

Light scattering arises from the variation of the dielectric properties, i.e. the refractive index. This is best explained by taking a look at a scheme of a typical single scattering event, shown in figure 1.6. The incident light with wave vector $\vec{k_0}$ is vertically polarised with respect to the scattering plane (which is in this figure the plane of the paper).

Applying Maxwell's equations to the vertically polarised electromagnetic wave propagating through a medium with the dielectric constant $\epsilon(\vec{r}, t)$, the amplitude of the electric field $E_S(\vec{R}',t)$ scattered in the scattering plane in direction of $\vec{R}'$ is given by [10]

$$E_S(\vec{R}',t) = -\frac{k^2 E_0}{4\pi}\frac{\exp[i(kR'-\omega t)]}{R'}\int_V \left[\frac{\epsilon(\vec{r},t)-\epsilon_0}{\epsilon_0}\right]\exp(-i\vec{q}\cdot\vec{r})\mathrm{d}^3\vec{r} \quad (1.31)$$

where V is the scattering volume, ϵ_0 the average dielectric constant of the medium and $\epsilon(\vec{r},t)$ the dielectric constant of the medium at a certain position $\vec{r}$ (relative to an arbitrary chosen point 0) at the time t. Equation (1.31) can be interpreted as the sum of the radiation of oscillating point dipoles. These dipole moments are induced in small volume elements by the incident electromagnetic field. The oscillation generates isotropically distributed light of the same frequency $f = c/\lambda$ as the incoming beam in form of a sphere like wave. Scattering thus is caused by fluctuations of the dielectric constant, because if $\epsilon(\vec{r},t) = \epsilon_0$, the amplitude of the scattered electric field is zero. A completely homogeneous medium does not scatter at all.

Static Light Scattering (SLS)

In static light scattering (SLS), the time averaged intensities are measured in the far field at different scattering angles θ. The intensity is then plotted against the scattering vector $\vec{q}$ to normalise for different wavelengths and refractive indices of solvent.

Being interested in colloidal suspensions and the gelation process, SLS can be treated for the special case of discrete scatterers. Equation (1.31) can be developed [11] for N particles with centres of mass at position $\vec{R}_j(t)$, where $\vec{r}_j(t)$ still describes a small volume element (interpreted as a dipole) in the particle j relative to its centre of mass

$$E_S(\vec{R}',t) = -E_0\frac{\exp[i(kR'-\omega t)]}{R'}\sum_{j=1}^N \Delta b_j(\vec{q},t)\exp[-i\vec{q}\cdot\vec{R}_j(t)] \quad (1.32)$$

The excess scattering length $\Delta b_j(\vec{q},t)$ of particle j is calculated by weighing the phase shifts of all the small scattering volumes of particle j with their excess scattering length density $\Delta\rho(\vec{r}_j,t)$ and integrating them over the volume

V_j of the particle j.

$$\Delta b_j(\vec{q}, t) = \int_{V_j} \Delta\rho(\vec{r}_j, t) \exp(-i\vec{q}\cdot\vec{r}_j) \mathrm{d}^3 r_j \qquad (1.33)$$

The excess scattering length density $\Delta\rho(\vec{r}_j, t)$, which can be regarded as the local excess density of the scattering material, is defined by

$$\Delta\rho(\vec{r}_j, t) = \frac{k^2}{4\pi}\left[\frac{\epsilon_p(\vec{r}_j, t) - \epsilon_s}{\epsilon_0}\right] \qquad (1.34)$$

where $\epsilon_p(\vec{r}_j, t)$ is the local dielectric constant of the particle, ϵ_s the dielectric constant of the solvent and ϵ_0 the average dielectric constant of the whole suspension. It is noteworthy that the excess scattering length $\Delta b_j(\vec{q}, t)$ is for identical particles related to the particle form factor $P(q)$ through

$$P(q) = \frac{\langle |\Delta b_j(\vec{q}, t)|^2 \rangle}{\langle |\Delta b_j(0, t)|^2 \rangle} \qquad (1.35)$$

where $\langle \rangle$ stands for the ensemble average.

Equation (1.32) consists like equation (1.31) of two terms: the first term corresponds to a spherical wave as a result of the superposition of individual spherical waves emitted by all scattering centers in the system. The second term contains the phase shift due to the interference between the scattering particles at positions $\vec{R}_j(t)$ in the sample.

However, the electric field is experimentally not easily accessible, but the scattered light intensity can be directly measured. Since the intensity is related to the field through $I(\vec{q}, t) = |E(\vec{q}, t)|^2$, it is possible to calculate from equation (1.32) the ensemble averaged scattered light intensity

$$\langle I_S(q) \rangle = \frac{E_0^2}{R'^2} \left\langle \sum_{j=1}^{N}\sum_{k=1}^{N} \Delta b_j(\vec{q}) b_k^*(\vec{q}) \exp\left[-i\vec{q}\cdot(\vec{R}_j - \vec{R}_k)\right]\right\rangle_e \qquad (1.36)$$

Please note that $\langle \rangle_e$ stands for the ensemble average, and all further angle brackets will denote ensemble averages unless noted otherwise. It is important to realise that in the case of a so-called ergodic system, the ensemble equals the time average, and in a static and dynamic light scattering experiment one normally measures the time rather than the ensemble average.

Exceptions are treated in chapter (1.2.2).

From now on, scattered fields and intensities will be regarded as a function of the scattering vector $\vec{q}$ instead of the position of the detector $\vec{R}'$. The two are equivalent because $\vec{R}'$ is in the direction of $\vec{k}_s$ and the latter is related to $\vec{q}$ by (1.13).

In the case of a dilute non-interacting system, the particles are uncorrelated, i.e. they behave like and ideal gas of colloidal particles. Equation (1.36) can then be rewritten as

$$\frac{d\sigma}{d\Omega}(\vec{q}) = \sum_{j=1}^{N} \left\langle |b_j(\vec{q})|^2 \right\rangle + \\ + \sum_{j\neq}^{N} \sum_{k=1}^{N} \left\langle b_j(\vec{q}) \exp(-i\vec{q}\cdot\vec{R}_j) \right\rangle \left\langle b_k^*(\vec{q}) \exp(i\vec{q}\cdot\vec{R}_k) \right\rangle \qquad (1.37)$$

where the second term is equal to zero because the particles are free to randomly explore the whole space over time, thus the exponential terms are distributed around 0 and cancel. This step explains why the time averaged intensity for non-interacting particles measured in SLS is the sum of the intensities scattered by each particle and contains information about size, structure and shape averaged for all particles.

Dynamic Light Scattering (DLS)

Dynamic light scattering (DLS) is also known as photon correlation spectroscopy (PCS). Where SLS uses a time averaged intensity to estimate the average static properties of the sample, DLS takes into account the temporal fluctuations of the scattering pattern. In colloidal suspensions, the so-called speckles move due to the Brownian motion of the particles in the solvent. As the fluctuations of the interference pattern represent these motions, dynamical properties of the system can be accessed via a time-resolved study of the scattered intensity. To do so, the same experimental setup as for SLS can be used, but the measured intensity is processed by calculating the intensity auto correlation function

$$\langle I(q,0)I(q,\tau)\rangle_t = \lim_{T\to\infty} \frac{1}{T} \int_0^T I(t)I(t+\tau)\mathrm{d}t \qquad (1.38)$$

The angle brackets $\langle\rangle_t$ denote a time average (which is the same as the ensemble average in case of an ergodic system), and the time difference between the two multiplied intensities is called lag time τ. For all starting times t and over all lag times τ, this function estimates the degree of correlation between two intensities at different values of τ. The first value of an auto correlation function is given by $\lim_{\tau\to 0}\langle I(0)I(\tau)\rangle = \langle I^2\rangle$, and for infinitely long lag times τ, its value drops to $\lim_{\tau\to\infty}\langle I(0)I(\tau)\rangle = \langle I\rangle^2$. The intensity auto correlation function is normalised to

$$g_2(q,\tau) = \frac{\langle I(q,0)I(q,\tau)\rangle}{\langle I(q,0)\rangle^2} \qquad (1.39)$$

In order to apply the theoretical model calculated from the Maxwell equations, we need to convert this into the electric field correlation function. The normalised field auto correlation function is defined by

$$g_1(q,\tau) = \frac{\langle E(q,0)E(q,\tau)\rangle}{\langle E(q,0)\rangle^2} \qquad (1.40)$$

and is related - under the assumption of a Gaussian intensity distribution of the beam profile - to the normalised intensity auto correlation function $g_2(q,\tau)$ through the Siegert relation

$$g_1(q,\tau) = \sqrt{g_2(q,\tau) - 1} \qquad (1.41)$$

The normalised field auto correlation function $g_1(q,\tau)$ approaches one at very short lag times τ (see figure 1.10), where the scattered field is highly correlated because the movement of one particle is too slow to cause a big relative displacement and therefore a big change in the scattered field. On the other hand, at very long lag times the correlation function must be zero as the particles are undergoing Brownian motion and their position and in consequence also their scattered fields are completely random compared to their initial positions - given a lag time big enough. The form and position

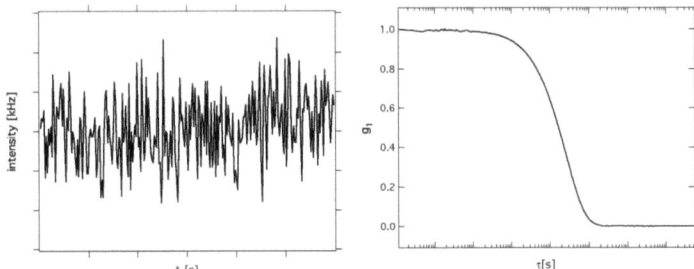

Figure 1.10: Dynamic Light Scattering on a dilute colloidal suspension of 19nm PS particles in a commercial standard setup. The fluctuations of the scattered intensity versus time are shown on the left, its normalised field auto correlation function $g_1(q,\tau)$ on the right.

of the decay allows for evaluation of the dynamic behavior of the scatterers in the system. Different quantities such as the particle size, polydispersity or viscosity of the solvent can be extracted from there. But in order to do so, a closer look at the calculated electric field is needed. The electric field is exactly the same as already mentioned for SLS (equation (1.32)). But as DLS treats normalised values, pre-factors can be omitted and the electric field can be described as

$$E_S(\vec{R}',t) = \sum_{j=1}^{N} \exp[-i\vec{q}\cdot\vec{R}_j(t)] \quad (1.42)$$

The right side of the equation above is essentially a sum of phase factors. Assuming that the particles are randomly distributed in the sample, the phase angles $(\vec{q}\cdot\vec{R}_j(t))$ are randomly distributed between 0 and 2π. The sum of all phase factors then can be regarded as a two-dimensional random walk of N vectors. The mean is zero because the random walk is symmetrical about its origin due to the randomly distributed phase angles, and with big N, $E_S(\vec{R}',t)$ becomes a complex variable with Gaussian probability distribution (following random walk theory using the Central Limit Theorem). The actually measured intensity $I(\vec{q},t) = |E(\vec{q},t)|^2$ is in consequence exponentially distributed and is visible in form of a speckle pattern, where small intensity maxima appear in front of mostly dark background. As the particle positions $\vec{R}_j(t)$ change due to Brownian motion, the phase factors change in time and the speckle pattern fluctuates. The speckle intensities can be measured and correlated, and through the Siegert relation the calculation of the normalised field auto correlation function is possible.

The normalised field auto correlation function $g_1(q,\tau)$ measured for dilute and identical particles is usually called the "measured intermediate scattering function" $f^M(q,\tau) = g_1(q,\tau)$ and simplifies for identical interacting spheres $(b_j(q,t) = b(q))$ to

$$f^M(q,\tau) = f(q,\tau) = \frac{F(q,\tau)}{S(q)} \qquad (1.43)$$

introducing the intermediate scattering function $F(q,\tau)$

$$F(q,\tau) = \frac{1}{N}\sum_j\sum_k \left\langle \exp\left\{-i\vec{q}\cdot\left[\vec{R}_j(0) - \vec{R}_k(\tau)\right]\right\}\right\rangle \qquad (1.44)$$

and defining the static structure factor $S(q)$ (see (1.27)) as

$$S(q) = F(q,0) \qquad (1.45)$$

For a dilute suspension of identical, non-interacting particles, the intermediate scattering function is given by

$$f(q,\tau) = \left\langle \exp\left\{-i\vec{q}\cdot[\vec{R}(0) - \vec{R}(\tau)]\right\}\right\rangle = \langle\exp[i\vec{q}\cdot\Delta\vec{r}(\tau)]\rangle \qquad (1.46)$$

where $\Delta\vec{r}(\tau)$ stands for the displacement of a particle and substitutes $(\vec{R}(\tau) - \vec{R}(0))$ in the second part of the equation. Therefore the intermediate scattering function which is measured by DLS contains information about the average motion of a particle. For an ergodic, dilute suspension of identical, non-interacting particles the displacement in Brownian motion is described by a Gaussian probability distribution. The field correlation function calculated from this results in

$$f(q,\tau) = \exp\left[-\frac{q^2}{6}\langle\Delta r^2(\tau)\rangle\right] \qquad (1.47)$$

with a particle mean square displacement of

$$\langle \Delta r^2(\tau) \rangle = 6D\tau \tag{1.48}$$

where the diffusion coefficient D for monodisperse spherical particles in free diffusion is given by the Stokes-Einstein relation

$$D_0 = \frac{k_B T}{6\pi \eta R_H} \tag{1.49}$$

with k_B being the Boltzmann constant, T the temperature, η the viscosity of the solvent and R_H the hydrodynamic radius of the particle. Knowing the temperature and the viscosity of the solvent, the calculation of the hydrodynamic radius of the particles is straight forward. The hydrodynamic radius for ideal spheres is related to the radius of gyration by $R_G = R_H \cdot \sqrt{0.6}$.

In more concentrated suspensions, the particles start to interact, and the diffusion coefficient in equation (1.48) has to take account of this. All direct interactions between the particles like for example Coulomb repulsion, Van der Waals attraction and excluded volume effects for hard spheres are already described in the static structure factor $S(\vec{q})$. Another important effect is the indirect hydrodynamic interaction $H(\vec{q})$; moving particles cause its surrounding liquid to flow, which will interfere with the motion of other particles. The effective diffusion coefficient for such cases is no longer independent of q and is given by

$$D_{eff}(\vec{q}) = D_0 \frac{H(\vec{q})}{S(\vec{q})} \tag{1.50}$$

DLS is mainly used for particle sizing, although there is another popular field of application; by adding a tracer of known size to a system, the viscosity can easily be measured with DLS.

The characteristic length scale probed by DLS is the same as the one for SLS; both methods are based on scattering caused by interference effects of singly scattered light. Therefore the investigated length scale is in the order of $\frac{2\pi}{q}$. But whereas SLS determines the radius of gyration R_G via the Guinier approximation and is limited to $R_G \gtrsim 15nm$ (below this size, the particle form factor does not change anymore for the accessible q-range), DLS has no

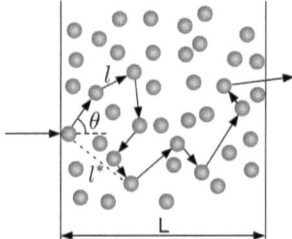

Figure 1.11: Photon path in a standard DWS transmission experiment. L is the sample thickness, l the scattering mean free path and l^* the transport mean free path.

lower theoretical limit for the measurable hydrodynamic radius R_H. Particles of any size which diffuse over the distance of the probed characteristic length scale create a constructive interference. Only the technical restriction of the shortest possible lag time τ and the fact that small particles scatter weakly set limits to the measurable particle sizes.

Diffusing Wave Spectroscopy (DWS)

In contrast to the single light scattering methods SLS and DLS, where multiply scattered light corrupts the signal and therefore needs to be avoided at any cost, Diffusing Wave Spectroscopy (DWS) takes advantage of multiply scattered light. DWS can only be applied to turbid media, as it depends on a high amount of scattering events experienced by a photon passing through a sample. A statistical approach - the diffusion equation - can then be taken to describe the photon transport in the sample [12]. DWS is the equivalent of DLS for turbid systems, but as the photon path is completely randomised, DWS measurements are not scattering angle dependent anymore. There exist only two different geometries: backscattering and transmission. Here, only transmission will be used and treated.

The number of scattering events of the photon on its way through the sample is a very important factor. For SLS and DLS, single scattering is required and the mean free path l of a photon should be longer than the sample thickness L ($l > L$). The mean free path is the average length between two scattering events and is given for dilute samples by

$$l = \frac{1}{\rho \sigma} \tag{1.51}$$

where ρ is the particle concentration and

$$\sigma = \frac{8\pi}{3}\left(\frac{2\pi}{\lambda}\right)^4 a^6 \left(\frac{m^2 - 1}{m^2 + 2}\right)^2 \tag{1.52}$$

is the total Rayleigh scattering cross section of the particle. $m = n_p/n_s$ is the ratio of the index of refraction of the particle n_p to that of the surrounding medium n_s.

The transport mean free path l^* on the other hand specifies the length scale on which a photon will loose any information about its initial direction. For distances larger than l^*, the light is therefore randomised, and the diffusion approximation can be applied. l^* is defined as

$$l^* = \frac{l}{\langle 1 - \cos(\theta) \rangle}, \quad l^* \geqslant l \tag{1.53}$$

The angle brackets denote an ensemble average over many scattering events, and θ is the scattering angle. l^* is the characteristic length scale for diffuse propagation of light, such as l for SLS and DLS. It sets the range of applicability for DWS which treats systems of high turbidity, starting at approximately $l^*/L \approx 5$ (for a further discussion, see 5.3.3).

A photon passing a turbid sample is scattered many times. All these scattering events have to be summed up to correctly describe the total scattering. First, the phase shifts of each scattering event are summed up to the total phase change

$$\Delta \phi_p(\tau) = \sum_{i=0}^{N} \vec{q_i} \Delta \vec{r_i}(\tau) \tag{1.54}$$

where $\vec{q_i} = \vec{k_i} - \vec{k_{i+1}}$ are the corresponding scattering vectors. With a high number N of scattering events, the Central Limit Theorem can be applied to the total phase change. It allows us to replace the phase change of each single scattering event with its average phase change. Assuming that the scattering events are uncorrelated and weighing the scattering vectors with the form factor of a single particle, the mean total phase change results in

$$\langle \Delta \phi_p^2(\tau) \rangle_t = \frac{2}{3} k_0^2 \langle \Delta r^2(\tau) \rangle_t \frac{s}{l^*} \quad (1.55)$$

where $\langle \rangle_t$ stands for a time average and introducing the path length $s = Nl$. As the mean total phase change depends only on the length of the scattering path, the total field correlation function can be calculated by summing up different paths weighed with the path-length distribution function P(s)

$$g_1(\tau) = \int_0^\infty P(s) \exp\left(-\frac{1}{3} k_0^2 \langle \Delta r^2(\tau) \rangle_t \frac{s}{l^*}\right) ds \quad (1.56)$$

The path-length distribution function P(s) is calculated using diffusion theory, where photon transport is described by

$$\frac{\partial U}{\partial t} = D_l \nabla^2 U \quad (1.57)$$

Here, $D_l = vl^*/3$ is the diffusion coefficient of light with v being the speed of light in the medium and U is the energy density of light. The diffusion approach allows for the determination of all the different scattering path lengths and the probability of a photon to take those paths. For the transmission geometry with an expanded incident beam (plane wave illumination) the field correlation can be calculated as follows [13]

$$g_1(\tau) = \frac{\left(\frac{L}{l^*} + \frac{4}{3}\right)\sqrt{\frac{6\tau}{\tau_0}}}{\left(1 + \frac{8\tau}{3\tau_0}\right) \sinh\left[\frac{L}{l^*}\sqrt{\frac{6\tau}{\tau_0}}\right] + \frac{4}{3}\sqrt{\frac{6\tau}{\tau_0}} \cosh\left[\frac{L}{l^*}\sqrt{\frac{6\tau}{\tau_0}}\right]} \quad (1.58)$$

$$\cong \frac{\left(\frac{L}{l^*} + \frac{4}{3}\right)\sqrt{\frac{6\tau}{\tau_0}}}{\sinh\left[\left(\frac{L}{l^*} + \frac{4}{3}\right)\sqrt{\frac{6\tau}{\tau_0}}\right]} \quad (1.59)$$

with $\tau_0 = 1/(k_0^2 D_0)$. This expression is very close to an exponential function, and therefore the following simplified expression is sometimes used

$$g_1(\tau) = \exp\left[-\frac{k_0^2 \left(\frac{L}{l^*}\right)^2}{6} \langle \Delta r^2(\tau) \rangle\right] \quad (1.60)$$

This function looks very similar to the result for DLS (see equation (1.47)). Both field correlation functions feature an exponential decay with $\langle \Delta r^2(\tau) \rangle$. But in the case of DWS, the decay depends on the initial wave number k_0 and L/l^* instead of the scattering vector q. This means the scattering angle is of no importance, but the transport mean free path has to be known. This is of great importance in particular for time evolving systems. l^* can be calculated using Mie theory for example, but it is usually measured to take into account all experimental conditions. Because the transmission T of a sample in the limit of strong multiple scattering is given by

$$T = \frac{5l^*/3L}{1+4l^*/3L} \sim l^* \tag{1.61}$$

it is proportional to l^*. By measuring the transmission of a sample and comparing it to the value of a system with known l^* measured under the same conditions, l^* can be obtained in a very easy way.

It is important to underline that DWS probes the motion of particles in the range of $\Delta r \approx \lambda_0 l^*/L$ and therefore is not comparable to DLS; the distance over which a particle has to diffuse to induce a constructive interference is noticeably smaller than the wavelength and can be as low as 1nm. This is caused by the fact that a photon passing the sample is scattered by a large number of particles, each of whom is in motion and therefore shifting the phase of the scattered light. As a consequence, the sum of the diffusion of all particles on the scattering path of the photon have to be in the order of λ.

Non ergodic Systems

Up to here, only ergodic systems have been discussed. In ergodic systems, the time average equals the ensemble average, so that a time averaged measurement contains the information about the ensemble average. This is the case for particles (and clusters) that undergo free diffusion. However, the situation changes completely for solid-like samples such as glasses or gels. Once the clusters connect to a space filling network (at the gel point), the individual particles are trapped and cannot explore the whole phase space anymore. Such a system shows non-ergodic behavior with a time average different from the ensemble average. A time averaged field correlation function of an ergodic system decays to zero, but the non ergodic decorrelates to a plateau because the particles cannot escape from their position in the network and therefore are always correlated to their position therein (see c) in figure 1.12). Furthermore, a series of time averaged measurements yields a set of non-reproducible correlation functions depending on the speckle which is observed (see a) in figure 1.12). The time averaged scattered intensity $\langle I \rangle_t$ varies for each measurement and is not equivalent to $\langle I \rangle_e$. The non-fluctuating part of the intensity contributes in DWS only to the background, therefore all time averaged correlation functions decay to zero. The plateau height of a non-ergodic system cannot be measured by a simple time averaged correlation function.

Different approaches to calculate the correct ensemble averaged g_1 have been proposed for DLS experiments [14, 15]. Pusey et al. [16] suggest to correct a time averaged field correlation function with the ensemble averaged intensity $\langle I \rangle_E$ obtained by measuring a rotating sample. Another, less elegant way is the brute force method, where a statistically significant number of measurements at different speckles, i.e. different sample positions, is mathematically averaged. Other techniques measure the real ensemble average by translating or rotating the sample slowly during data acquisition, adding a second, slow decay to the field correlation function. All these approaches have been developed for single light scattering, and some can be extended to multiple light scattering. But in each method, the sample needs to be moved physically, and the measurements are rather long. For these reasons, the techniques mentioned above are not very well adapted for time resolved measurements and fragile structures.

A CCD camera is able to measure many correlation functions at the same time, which than can be averaged in so-called multi speckle correlation measurements. But due to the slow response times of the camera, this method

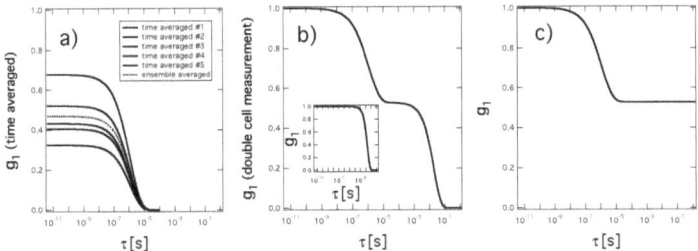

Figure 1.12: Illustration of field correlation functions g_1 for non ergodic systems. Five time averaged measurements are shown in a), with the plateau height interpreted as baseline. Dotted is the corresponding ensemble and time averaged correlation function. The measured g_1 using the double cell technique is plotted in b); the inset is g_1 of the second cell alone. The real ensemble averaged field correlation function which can be restored from both measurements a) and b) is shown in c).

is not well suited for our samples neither.

Double cell technique Romer et al. [17] and Scheffold et al. [18] present another way to measure ensemble averaged field correlation functions. This method is called double cell technique because attached to the cell with the investigated sample is another cell containing an ergodic turbid medium (see figure 1.13). A laser beam passes first through the non ergodic sample and then through the ergodic medium. The sample features a faster decay which stops at a certain plateau height. The second, ergodic medium is tuned in such a way that its dynamics is at least an order of magnitude slower than the first decay of the non ergodic sample. The second cell then induces a full (ensemble averaged) decorrelation at high lag times. It is shown in [18] that the resulting correlation function for weak absorption and $L \gg l^*$ is then given by a simple multiplication rule (for the simplified case of independent scattering in both cells which means no loop-like photons paths between both

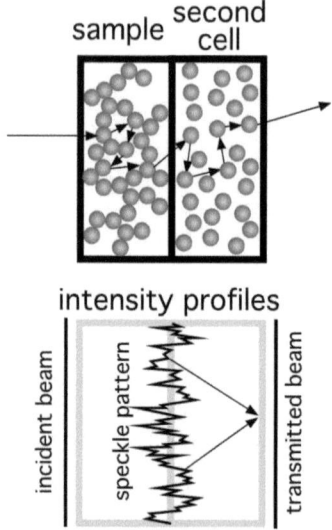

Figure 1.13: Scheme of the double cell setup. Top: the incident beam passes first the (non ergodic) sample under study and then a second cell which contains an ergodic system. The slow motion of the scatterers in the second cell gently shakes and randomizes the speckle pattern of the non ergodic medium and ensures proper ensemble averaging of the light scattered by the first cell. Bottom: the intensity profile of the incident beam is homogeneous, whereas the non ergodic sample produces a time-independent speckle pattern. These speckles can be considered as point sources of light which are multiply scattered by the second cell and therefore washed out. The total transmitted beam features an averaged intensity of the speckle pattern (indicated by the two arrows).

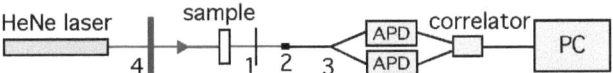

Figure 1.14: Scheme of the Two-cell Echo approach. Besides the rotating diffuser (4) between laser and sample it is identical to a standard DWS setup as seen in figure 3.1. Details are given in the text resp. in section 3.1 where the standard DWS setup is described.

cells)

$$g_{2\ total} - 1 = (g_{2\ sample} - 1) \cdot (g_{2\ scndcell} - 1) \qquad (1.62)$$

The total field correlation function $g_{2\ total}$ thus features a two step decay (see b) in figure 1.12). Dividing it by the separately measured field correlation function $g_{2\ scndcell}$ of the second cell only (using equation (1.62)), the real ensemble averaged field correlation function $g_{2\ sample}$ of the sample remains. This method has the advantage to be very simple and fast, and furthermore the sample doesn't have to be moved physically. But it can only be applied to turbid media which means for DWS.

Two-cell Echo approach The so-called Two-cell Echo approach proposed by Zakharov et al. [19] expands the second cell DWS scheme using the Echo technique developed by Pham et al. [20] and allows to determine the auto correlation function for $0.1s < \tau < 450s$ within a measurement of the length of only 450 seconds. The setup as seen in figure 1.14 is identical to a standard DWS experiment like in figure 3.1 except for the fact that a rotating ground glass i.e. diffuser (4) is placed between the incoming laser beam and the sample. 1, 2 and 3 designate a horizontal polariser, colllimator and single mode fiber beam splitter respectively. The diffuser rotates at high speed (up to 75 Hz) and it is therefore possible to measure the "echos" of the correlation function i.e. the repetition of the correlation peak after one full revolution of the diffuser (as the diffuser is a static scatterer, the same speckles are

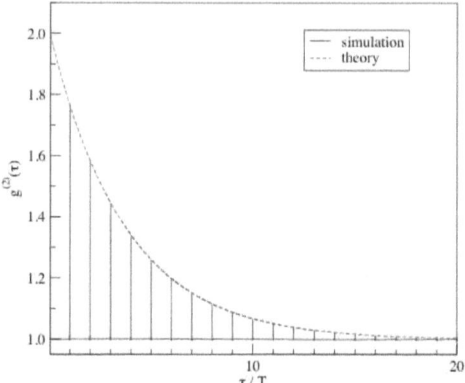

Figure 1.15: Simulated echos of a fluid (continuous lines) and the predicted dynamics from its Brownian motion (dashed line) showing that the echoes follow the dynamics of the sample (the x axis is the lag time τ divided with the period T of the diffuser).[6] The echos of a rigid sample i.e. static scatterer would show a constant intercept of 2, all echos would have identical heights.

generated after one revolution). These echos of the correlation function follow the samples properly ensemble averaged dynamics [20] (see figure 1.15), so the sample dynamics can be obtained by measuring the temporal evolution of the echos.

Each echo signal is generated by a large number of independent speckles thus efficient ensemble averaging is performed. The scattered intensity is measured with a photon counter instead a correlator, saved as a photon trace and treated afterwards with software which correlates the echo signals. This method is ideally suited to investigate the expected glass phase in the gelation process as very slow processes at long lag times τ can be characterised in

---

[6]Reused with permission from K. N. Pham, S. U. Egelhaaf, A. Moussad, and P. N. Pusey, Review of Scientific Instruments, 75, 2419 (2004). Copyright 2004, American Institute of Physics.

a short measurement time, allowing for time resolved, ensemble averaged measurements.

Data analysis Once the correct ensemble averaged field correlation function is acquired, the data can be analysed. Krall and Weitz developed a theory for the interpretation of $\langle \Delta r^2(\tau) \rangle$ for fractal particle gels [5, 21]. As observed by DLS, a gel is regarded as an object assembled from fractal clusters of size $R_{c,critical}$, so-called "blob". These blobs again are assembled from smaller blobs, and so forth down to the size of a single particle. All blobs and particles are bound in the network and thus localised, but they still can move around their average position. The blob's motion is coupled to the motion of its smaller blobs, each of them having their own collective motion called mode. Each mode can be treated as an independent, overdamped harmonic oscillator characterised by a spring constant and a relaxation time. The motion of the gel (the biggest blob) is then given by the integral over all modes for all possible blob sizes, resulting in

$$\langle \Delta r^2(\tau) \rangle = \delta^2 \left[1 - e^{-\left(\frac{\tau}{\tau_\delta}\right)^p} \right] \quad (1.63)$$

The cross-over time τ_δ is a parameter representing the characteristic time scale at which the motion of the particles starts to be limited by the network. It is related to the characteristic fluctuation of the cluster τ_c by $\tau_\delta = 0.35\tau_c$. The plateau height stands for the maximum mean square displacement δ^2 of the gel. The motion of a gel is sub-diffusive, which means basically that the retaining force is not linear but increasing with elongation from the initial position. This is represented by the exponent p which can be read out from the initial slope in figure 1.16 on the right side. In a free diffusive motion, $\langle \Delta r^2(\tau) \rangle$ follows a power law with exponent one, a gel according to the Krall/Weitz model exhibits an exponent of 0.7. While equation (1.63) describes the microscopic state of the system, its parameters are linked to macroscopic properties of the system through the model of overdamped os-

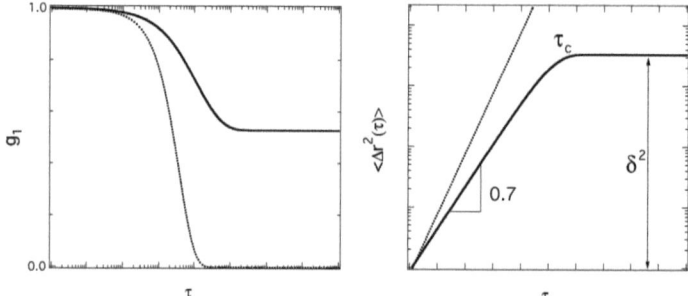

Figure 1.16: Schematic representation of the field correlation function g_1 obtained for a particle gel (left) and its corresponding mean square displacement $\langle \Delta r^2(\tau) \rangle$ (right). Dotted line: the ergodic case for comparison. Details in the text.

cillators. The elastic modulus G_0 is then given by

$$G_0 = \frac{6\pi\eta \cdot 0.35}{\tau_\delta} \qquad (1.64)$$

whereas the cluster radius R_c can be calculated with

$$R_c = \frac{2d_f D_0}{(2+d_b)} \frac{a\tau_\delta}{0.35 \cdot \delta^2} \qquad (1.65)$$

where d_b is the bond dimension with an estimated value of 1.1 (by computer simulations) and a the particle radius.

For the investigation of turbid gels, the Krall/Weitz model can be applied to DWS theory by combining equations (1.63) and (1.58) to

$$g_1(q,\tau) = \frac{\left(\frac{L}{l^*} + \frac{4}{3}\right)\sqrt{k_0\delta^2\left(1 - e^{-\left(\frac{\tau}{\tau_\delta}\right)^p}\right)}}{\sinh\left[\left(\frac{L}{l^*} + \frac{4}{3}\right)\sqrt{k_0\delta^2\left(1 - e^{-\left(\frac{\tau}{\tau_\delta}\right)^p}\right)}\right]} \qquad (1.66)$$

But it remains to be seen that the Krall/Weitz model can be directly applied to the interpretation of DWS experiments with such small particles as we used. This issue will be treated in the forthcoming subsection 5.3.1.

1.2.3 Neutron scattering

The basic principles for neutron scattering are the same as already described in the preceding chapters. The fundamental difference between neutrons and photons is that neutrons scatter at the nuclei of atoms and the photons scatter due to the polarisability of a particle. As neutrons are weakly interacting with matter, problems with multiple scattering are less important and can be avoided by adjusting the scattering cross section (index matching or contrast variation, see below). Due to the fundamental difference of the scattering mechanism, samples appearing completely turbid in visible light can be transparent for neutrons and vice versa.

The differential cross section for neutron scattering is given by [22]

$$\frac{d\sigma}{d\Omega}(\vec{q}) = \langle b \rangle^2 \sum_{j,k=1}^{N} \langle e^{i\vec{q}\cdot\vec{r}_{jk}} \rangle + N \left(\langle b^2 \rangle - \langle b \rangle^2 \right) \tag{1.67}$$

where the first term represents the coherent scattering part and contains all information about the structure of the particles and their spatial arrangement. Coherent scattering is caused by the interference of the scattering of all nuclei which all have the same average scattering length $\langle b \rangle$, and its scattering cross section is given by $\sigma_{coh} = 4\pi \langle b \rangle^2$. The second term corresponds to the incoherent scattering part, where no interference terms appear. Hence, this part bears no information about the structure of the sample and contributes only to a q-independent background intensity. The incoherent scattering cross section is given by $\sigma_{incoh} = 4\pi(\langle b^2 \rangle - \langle b \rangle^2)$.

The scattering lengths of different nuclei vary strongly and unsystematically for neutrons. Furthermore, big differences exist between the scattering lengths of different isotopes (see table 1.1).

This allows for the so-called contrast variation, where the scattering contrast i.e. the excess scattering length density $\Delta \rho$ of the investigated sample

Isotope	b [10^{-15}m]	σ_{coh} [barn]	σ_{incoh} [barn]
^1H	-3.74	1.76	80.27
^2H	6.67	5.59	2.05
O	5.803	4.23	0.0008
C	6.65	5.55	0.001
S	2.85	1.02	0.007

Table 1.1: Examples for the coherent scattering length b and the coherent resp. incoherent cross-sections σ_{coh} and σ_{incoh} of several nuclei

can be manipulated. Chemically identical isotopes can be mixed or replaced without altering the sample characteristics, but changing the neutron contrast drastically. By the choice of the appropriate mixture of isotopes, a contrast can be created (or diminished) between for example particles and solvent or different parts of particles. For example, the big difference of the scattering length between $b_H = -3.74$ and $b_D = 6.67 \cdot 10^{-15}m$ results in scattering length densities ρ for H_2O and D_2O (heavy water) of -0.56 resp. 6.33 $\cdot 10^{-6}$ Å^{-2}. By the substitution of H with D in the particle or H_2O with D_2O in the solvent, the excess scattering length density of the sample can be varied in a wide range from negative to positive values, including the index matching point where $\Delta\rho = 0$.

Neutrons have completely different wavelengths than photons; cold neutrons typically used in SANS experiments have wavelengths in the order of $0.3nm < \lambda < 3nm$. The length scales probed in a measurement in single scattering condition are in the order of $\frac{2\pi}{q} = \frac{\lambda}{2\sin(\theta/2)}$, so much smaller wavelengths probe much smaller length scales. To resolve large structures, very small scattering angles have to be chosen for neutrons.

Chapter 2

System

In previous light and neutron scattering experiments on aggregation and gelation of colloidal suspensions, different model systems such as polystyrene spheres, silica, PMMA and gold [2, 21, 23] have been studied. The idea here is to develop further a work already presented by Romer et al. [23], where colloidal suspensions of polystyrene have been destabilised by the urea/urease process (see 2.2) and measured using simultaneous time-resolved DWS and SANS experiments with an older version of the setup described in chapter 3.3.

2.1 Choice of System

Romer et al. used in their work charge stabilised colloidal suspensions of polystyrene particles of a diameter of 190 nm. Being interested in the fractal structure of the clusters and the final gel, small particles seem to be the best choice as the SANS I instrument at the PSI used for these experiments has a limited q-range of $6 \times 10^{-3} nm^{-1} < q < 10.5 nm^{-1}$. Moreover, the fractal region for a particle gel is limited to the region $R_{c,critical} > \frac{2\pi}{q} > a$. The use of small particles thus allows us to enlarge this fractal regime and to shift it into the accessible q-window of SANS I, and we can observe important properties such as $R_{c,critical}$, d_f and local correlation effects between neighbouring

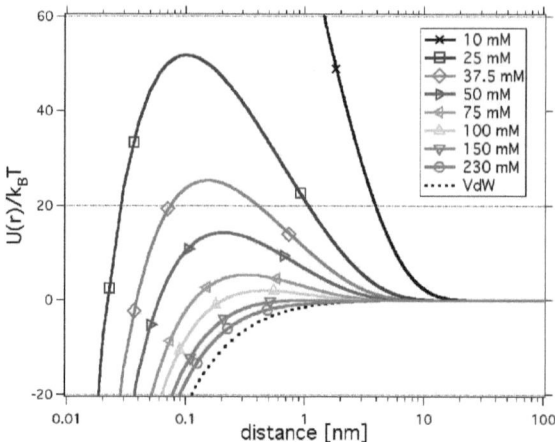

Figure 2.1: DLVO potentials calculated for a 19nm PS colloidal suspension in dependance of the amount of added salt.

particles (see figure 1.9). The smallest commercially available polystyrene $((C_8H_8)x)$ spheres have a diameter of 19 nm with a polydispersity of 16.3% (Interfacial Dynamics Corporation, Sulfate White Polystyrene Latex, product nr. 1-20). They feature a surface charge density of $1.8 \mu C/cm^2$, and the corresponding calculated DLVO potentials in dependance of the amount of added salt are shown in figure 2.1. Due to the small size of the particles, the DLVO potential features no visible secondary minimum. The critical coagulation concentration ccc is reached at approximately 150 mM.

At volume fractions of a few percent, the resulting suspensions are nearly transparent (yellowish suspension, left side of figure 2.2). However, the sample appearance changes drastically upon aggregation and the resulting suspensions and gels become turbid ($l^* \approx 400 \mu m$ for the final gel at $\phi = 3.8\%$, right side of figure 2.2), so DWS measurements become feasible after a minimal size of the aggregating clusters.

Figure 2.2: Colloidal suspension of 19nm PS spheres (diameter) with a volume fraction of 3.8 volume percent in a buoyancy matched mixture of H_2O/D_2O. Left: stable suspension, slightly yellowish but transparent. Right: gelled out sample, turbid. Both cells are 2mm thick.

In order to be able to follow the aggregation and gelation process in a time-resolved manner, the individual measurements should be significantly shorter than the gelation process. To reach a statistically significant amount of at least 100'000 coherently scattered neutrons, a typical SANS measurement requires around 15 minutes. The aggregation therefore needs to be slow (around 3 hours to the point where it gets non ergodic), and the PS particles have to be buoyancy matched in order not to sediment during this long process. Buoyancy matching is achieved by mixing H_2O and D_2O in the ratio 52:48 by volume, as polystyrene has a density of $\sigma = 1.055 g/cm^3$ which lies between $\sigma_{H_2O} = 0.998 g/cm^3$ and $\sigma_{D_2O} = 1.105 g/cm^3$ (all at 20 °C). The dilution of H_2O with D_2O in the system results also in an advantageous scattering cross section for neutron scattering which is relatively close to the contrast match point of polystyrene spheres and thus helps to avoid multiple scattering effects.

2.2 Method of destabilisation (urea/urease)

To induce aggregation and gelation in the colloidal suspension mentioned above, additional ions have to screen the particle charges (see 1.1.2). As we work at relatively high volume fractions of a few percent ($0.01 < \phi < 0.18$), the simple addition of salt by mixing with a second solution is not applicable. The resulting gradients of the ion concentration would disturb noticeably the aggregation process, causing inhomogeneous cluster growth. Moreover, the speed of the aggregation process should be tunable to allow for time-resolved measurements during the gelation. Both problems are overcome by using a special method: the urea/urease process.

$$(NH_2)_2CO + 2H_2O \xrightarrow{Urease} CO_3^{-2} + 2NH_4^+$$

The hydrolysis of urea is an enzymatic process where urease is the catalyst. This method was developed in material science to enhance ceramic processing [24]. Urea dissolved in water is hydrolysed to carbonate and ammonium. This process has the advantage to produce ions homogeneously in situ. In addition, the kinetics can be easily controlled: the amount of urea defines the final ion concentration and the amount of urease sets the speed of the nearly linear ion production. Temperature and pH influence the kinetics of the reaction as well. In our case, typical values of 4.54 weight percent of urea (Fluka Urea puriss. p.a. 51459) and 20 units/ml of urease (Roche Diagnostics: Urease, Lyo., SQ) have been used. At a temperature of 25 °C, this results in a nearly constant ion production during 280 minutes and a final ion concentration of 227 mMol/l as shown in figure 2.3. As the ion concentration increases constantly, the potential barrier of the charged particles decreases steadily. The particle aggregation therefore starts as a RLCA process with growing sticking probability until the potential barrier vanishes completely and DLCA sets in at 125 minutes (the ccc for our conditions is reached at 150 mM). This crossover from reaction to diffusion limited aggregation complicates the quantitative description of the gelation, but the advantages of

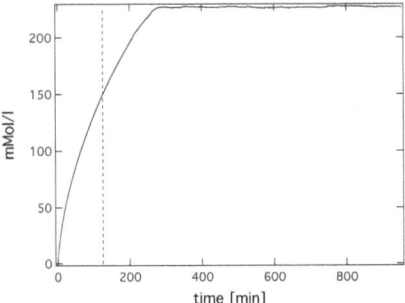

Figure 2.3: Temporal evolution of the ion concentration using the urea/urease process. 4.54% urea are solved in a buoyancy matched mixture of H_2O and D_2O, and 20 u/ml urease are added. The dotted line marks the ccc of 150 mM at 125 minutes.

the urea-urease method outweigh this problem.

Sample preparation is quite easy; first, the correct amount of urea is dissolved in the sample at the preferred temperature. Then a urease solution solved in the same solvent as the sample is added (ten percent of the sample volume). As urease is active at temperatures above 4 °C and its lifetime is below 24 hours, it has to be prepared under controlled conditions. Crystalline urease has to be stored at temperatures between 2-8 °C and needs to be dissolved using an ice bath but then can be frozen for storage. The frozen urease is molten and added to the sample, where the immediate temperature increase activates the enzyme.

2.3 Sample preparation for the investigation of time resolved aggregation and gelation

All data presented in this chapter were measured under the same conditions. The sample was thermally stabilised at 25 °C at ambient pressure. The 19nm

PS sample was prepared and mixed in a vial before filled into round quartz cuvettes of 19mm inner diameter and 2mm sample thickness (Hellma 120-QS). The volume fraction of PS is $\phi = 0.038$, the solvent is a mixture of H_2O and D_2O in the volume ratio 52/48 and contains 4.54 weight percent of urea and 20 u/ml urease.

Chapter 3

Setup

3.1 DWS

Standard DWS setups are relatively uncomplicated and compact enough to be mounted within a few hours, including alignment and testing. The scheme of a simple setup is shown in figure 3.1. A small HeNe laser (Uniphase 1145P, 30mW) provides enough light to probe a sample. The scattered light is coupled with a collimator (2) (Schäfter+Kirchhoff 60FC-0-M8-33) into a single mode fiber (3) (Schäfter+Kirchhoff FSB-630-Y). This special fiber designed and manufactured for our needs by Schäfter+Kirchhoff will further on be referred to as "single mode fiber beam splitter". One end features a common single mode fiber with a standard FC/PC connector, where the light is coupled in through the collimator. In the middle, a second single mode fiber is fused to the first one so that the light will be split between the two of them with a ratio of 50:50. These two fibers then end as well in FC/PC connectors, which will feed the light into two Avalanche Photo Diodes (APD, Perkin Elmer SPCM-AQR-13-FC). To filter non-diffuse light, a horizontal polariser (1) is placed in front of the collimator. Both signals are then cross correlated with an external USB correlator (correlator.com Flex99R12FCS) and read out by a laptop PC. The cross correlation of the two detectors measuring identical light intensities is a technical trick called pseudo cross correlation, it

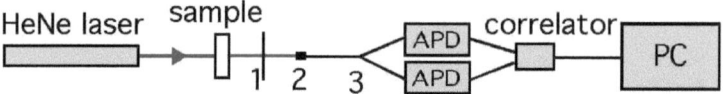

Figure 3.1: Scheme of a simple DWS setup in transmission geometry.

reduces signal alterations through detector specific properties (after pulsing). Using these components, correlation lag times of $12.5ns < \tau < 3436s$ can be accessed, as the correlator uses a multi-tau scheme. The lag times are logarithmically distributed, resulting in a heavy ponderation of short lag times (more points there). To ensure good data quality, time averaging should be performed over roughly 1000 times the slowest relaxation time or more. But the measurements should also be short enough to capture the time evolution of the sample. A good compromise is in our case a measurement length of 900s.

3.1.1 Non ergodic DWS

The two-cell DWS setup described in chapter 1.2.2 for non ergodic measurements has been further developed. The second cell containing an ergodic colloidal suspension to achieve a real ensemble average is now replaced by a moving static scatterer. In a first step, a piece of sandblasted glass was mounted on a motorised translation stage (Physik Instrumente M-150.10 with DC motor C-120.80 and motor controller card C-832 with the corresponding software). This construction is called "translational second cell" see figure 3.2. As the step resolution is very high (0.06 μm) and the travel of the stage long (50 mm), the slow and consistent translation that results in the typically needed correlation decay can be carried out for around 15 minutes before the translation stage reaches its end position. This time is sufficiently long for a typical DWS measurement. The advantages of this approach are: the characteristic decay time τ_{char} of the second decay can easily be tuned in

Figure 3.2: Picture of the "translational second cell": a static scatterer (a piece of sandblasted glass) is moved on a high precision translation stage.

the range of $0.001s \lesssim \tau_{char} \lesssim 50s$ by varying the translation speed instead of diluting a suspension (or mixing it with a viscous liquid) and the second cell doesn't need to be thermally stabilised. Furthermore, the transmission is higher and the correlation function is smoother i.e. a larger part of the correlation function can be evaluated after the division by the second decay. Another advantage is that the second cell is separated by a distance of several centimeters from the sample, reducing the possibility of loop like photon scattering paths. The scattered light from the sample has to be focussed to the ground glass with a lens to achieve high enough photon count rates.

To improve the second cell further, a high resolution step motor (MAE HS200-2216-0100-AX08) is used to rotate a sandblasted piece of glass. This construction is called "rotational second cell" (see figure 3.3). The control and power unit contains a programmable stand-alone step motor controller (IMS MX-CS101-400 Microlynx4), the motor can be turned on and off by switches, and the rotation speed can be selected from three preprogrammed values. In combination with a gear transmission ratio of 436, each revolution is split up in 22'323'200 steps which results in a step size of around 10 nm for a typical measurement. The rotating unit is constructed such that the rotating ground glass is fixed in the middle of a rotating ring. Depending on

Figure 3.3: Picture of the most recent version of the second cell: a rotating piece of ground glass, set in motion by a step motor. The stand-alone programmable power and control unit is on the right side of the picture.

the distance between collimator and the rotation axis (0 - 40 mm), the speed of the static scatterers vary and therefore the correlation time. In combination with the variable rotation speed, a wide range of correlation times can be easily accessed ($0.001s \lesssim \tau_{char} \lesssim 500s$). Larger distances from collimator to rotation axis are preferable because more speckles are passed through per motor step, resulting in a more continuous and smoother movement. Moreover, a higher total speckle number per revolution ensures better statistical averaging (as the ground glass is a static scatterer, its speckle pattern repeats with each revolution). The compact design and the stand-alone controller (instead of an additional computer for the motor controller card like in the case of the translation stage) make this second cell setup easy to transport and to incorporate in other experimental setups, which is required for the use in the combined SANS/DWS setup. Compared to the translation stage it provides also a higher step size resolution, a wider range of accessible decorrelation times and a continuous motion which makes very long acquisition times possible.

Moreover, the correlation function of the rotational second cell only is for

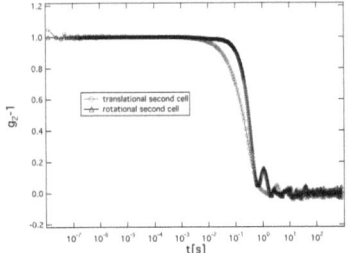

Figure 3.4: Normalised auto correlation $g_2 - 1$ of the two second cells (translational and rotational). Both measurements were 900s long and have been done by replacing the sample cell with a static scatterer, so that the resulting decay shows the de-correlation of the second cell only.

our purposes better than the translational one. Measurements of both second cells are shown in figure 3.4; in both cases, the sample cell was replaced with a static scatterer in order to measure the de-correlation of the second cell only, and both measurements were 900s long. The speed of the second cells was adjusted in such a way that the base line length was approximately three orders of magnitude (the rule of thumb for reasonable data quality). It is clearly visible that the rotational second cell provides a correlation function where the de-correlation sets in later and with a steeper decay. The small peak at the end of the decay of the correlation function for the rotational second cell is perfectly reproducible and is caused by vibrations of the sample table. As the data of a measurement is first divided by the second cell and subsequently analysed for the first decay only (which contains all accessible information about the sample), the file is only analysed up to lag times around the beginning of the decay of the second cell. The small vibrational peak and baseline fluctuations are out of this range and therefore of no significance. But due to the later onset of the decay of the rotational cell, the first decay can be analysed up to five time longer lag times compared the data measured with the translational cell (up to 0.075s instead of 0.015s in

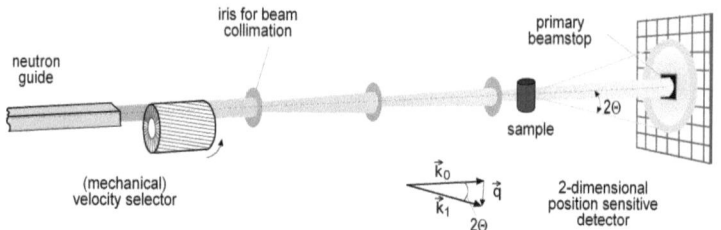

Figure 3.5: Scheme of the SANS-I instrument at PSI[7].

the case above).

3.2 SANS

Our interest lies in the investigation of colloidal suspensions and gels on a large range of length and time scales using light and neutron scattering. The scattering angle for neutrons must be very small to compensate for the short wavelength of the neutrons in comparison to visible light, so we use Small Angle Neutron Scattering (SANS). All our experiments have been conducted at the SANS-I device [25] at the Swiss neutron source at the Paul Scherrer Institut (Villigen, Switzerland, http://sinq.web.psi.ch).

As illustrated in figure 3.5, the (cold) neutrons from a spallation source exit the neutron guide and pass a monochromator (a helical slot velocity selector) which selects the neutron wavelength in the range of $0.45 nm < \lambda < 4 nm$ with $\delta\lambda \cong 10\%$. A collimator of variable length between 1m and 18m collimates the neutron beam prior to the sample. The sample is situated in a variable sample environment (see figure 3.6) where different devices can be installed. Standard options include automated sample holders on a xyz-translation table, vacuum chambers, temperature chambers and electromagnetic cells. The scattered neutrons then enter a vacuum tank

---

[7]Copyright (2005) by J. Kohlbrecher, Paul Scherrer Institut, Switzerland

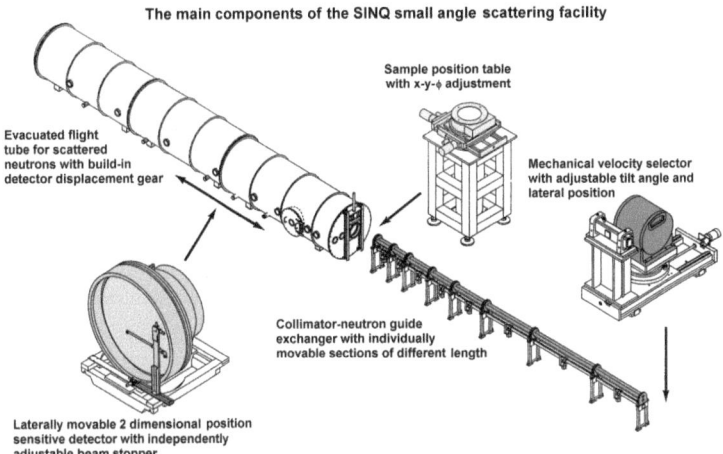

Figure 3.6: Main components of the SANS-I instrument at PSI[8].

in which a 2D ^{3}He detector can be driven to distances between 1m and 20m. The detector consists of a matrix of 128×128 detectors with a size of $7.5 \times 7.5 mm^2$ and can also be laterally displaced up to 0.5m. The accessible q-range is $6 \times 10^{-3} nm^{-1} < q < 10.5 nm^{-1}$. To enhance intensity, enlarge the q-range and improve the instrument resolution, biconcave magnesium flouride (MgF$_2$) lenses can be used to focus the neutrons [26, 27].

3.3 combined setup

The free space on one of the variable sample environments of the SANS instrument described in the chapter before was used to mount a light scattering setup similar to the one described in chapter 3.1. A scheme of the combined setup is shown in figure 3.7, and two pictures are shown in figure 3.8. This setup allows for measurements of DLS, DWS and non ergodic DWS simul-

---

[8]Copyright (2005) by J. Kohlbrecher, Paul Scherrer Institut, Switzerland

taneously to the SANS measurements. A thermally stabilised sample holder with 16 sample chambers on a xyz-translation table was provided by the PSI, leaving enough free space to mount the components for the light scattering. To enhance the q-range, 10 biconcave MgF_2 lenses focus the neutron beam to the sample. The vertically polarised laser beam emitted from a HeNe laser is reflected on a silicon waver. This silicon waver features an aluminum coated surface which makes it reflective for visible light, but still remains transparent for neutrons. The single mode fiber beam splitter is mounted first on a collimator which is aligned for a specific angle and which collimates the scattered intensity directly from the sample (marked as DLS in figure 3.7). For transparent samples, DLS measurements can be carried out at this specific angle of 32° (tests and calibration are done with reference systems).

Once the sample is too turbid for DLS due to multiple scattering, the single mode fiber beam splitter is mounted on the other collimator which is coupling in the light from the second cell for non ergodic DWS measurements. The second cell is a piece of rotating ground glass which accomplishes a true ensemble average (see chapter 3.1.1) of the scattered light of the sample which is focussed onto the ground glass through a lens. Both collimators (for SLS and for DWS) feature a standard FC/PC fiber connection and keep their aligned position albeit the connection and de-connection of a fiber.

3.3.1 Experimental settings used in the aggregation and gelation experiments

DLS correlation functions have been measured for 120 seconds at the scattering angle $\theta = 32°$, accessing lag times in the range of $12.5ns \leqslant \tau \leqslant 114s$. DWS measurements have been taken during 900 seconds in transmission geometry using the rotational second cell and a horizontal polariser (photons scattered once or only a few times basically keep their initial vertical polarisation and are filtered out). The second cell is rotating during all measurements, and its slow decay requires the long measurement times of 900s for a good base line quality of the correlation function (lag times of

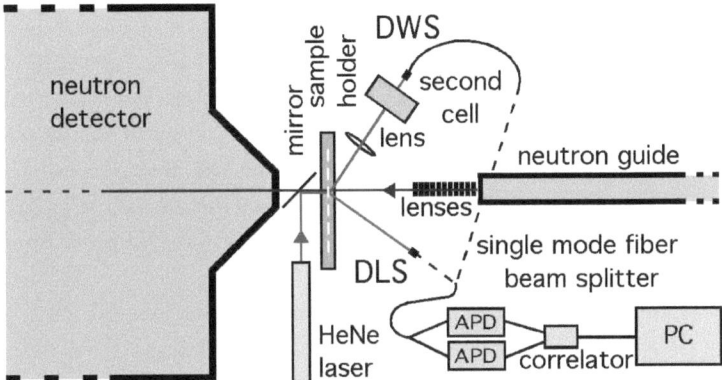

Figure 3.7: Scheme of the combined setup for simultaneous neutron and light scattering. The SANS-I instrument at PSI was extended with a light scattering setup allowing for DWS (both ergodic and non ergodic) and DLS.

Figure 3.8: Pictures of the setup. Left: top view, right: side view.

$12.5ns \leqslant \tau \leqslant 859s$).

The used SANS setting included 10 MgF$_2$ lenses and used a round aperture of 16mm diameter. The monochromator was set to a wavelength of $\lambda = 12.67nm$, the collimator length was 18m and the detector distance 20.3m. The detector was positioned with a lateral offset of 300mm to increase the q range to $0.008nm^{-1} \leqslant q \leqslant 0.215nm^{-1}$. A standard measurement took typically 900 seconds. The raw data has been corrected for suspension and cell background scattering and normalised with a reference measurement of H_2O (1mm cell thickness).

The sample preparation and the cuvette are described in section 2.3.

Chapter 4

Aggregation, cluster formation and sol-gel transition in a moderately concentrated colloidal suspension

The aggregation and gelation of colloidal suspensions has already been studied for a wide range of samples, and our chosen system (see section 2.1) is not to be examined for the first time. It is a standard model system chosen for its convenient particle size which allows us to examine a fractal gel using the previously described techniques DLS, DWS and SANS. The particularly interesting points in this study are the range of the examined sample volume fractions, the slow rate of the gelation process due to the controlled in-situ ion production using urea and urease and the use of a combined DLS/DWS/SANS setup. In combination, these three points allow us to investigate the aggregation and gelation process of a model system with a spatial and dynamic resolution that provides new information and insights into the sol-gel transition in colloidal suspensions.

Dynamic and static properties of gels and the different stages of a suspension undergoing a gelation process have already been (separately) measured for dilute [5] and concentrated [17, 28] samples, however a number of open questions remained. Using the DWS technique, Bissig [29] investigated the dynamics of colloidal aggregation and gelation at a moderate volume fraction of ϕ=0.045 and observed a sol-gel transition which did not exhibit the standard RLCA/DLCA characteristics everywhere. He speculated on the existence of a glassy phase just before the space filling network is spanned, in accordance with a precedent work of Segre et al. [30] who measured static and dynamic properties of moderately concentrated samples. Time-resolved measurements of the system investigated by Bissig have not been possible because the gelation process takes only 7 minutes. The kinetics of the gelation process in our system is chosen such that it takes roughly 350 minutes which allows us to follow the temporal evolution of the process with sufficient accuracy. The use of parallel dynamic (DLS/DWS) and static (SANS) measurements delivers complementary data sets measured so far only for concentrated systems by Romer et al. [23, 31, 32].

4.1 Structure S(q)

The radially averaged data of a sample prepared as described in section 2.3 measured using the settings mentioned in subsection 3.3.1 is shown in figure 4.1. The structure of our sample varies strongly during the aggregation and gelation process and can be split up in four phases I-IV.

At the starting point (phase I), the suspension was measured in its initial state before the addition of urease, shown in the lowest curve. The structure peak of a highly correlated liquid caused by the high particle charge is clearly visible in this first phase.

The second phase (II) starts with the production of ions, when the screening of the particle charges sets in. As the potential barrier of the DLVO potential decreases, the suspension looses its correlation and the corresponding structure peak disappears, accompanied by an increase of the forward scattering (increasing intensity in the low q regime). The curve 25 minutes after the

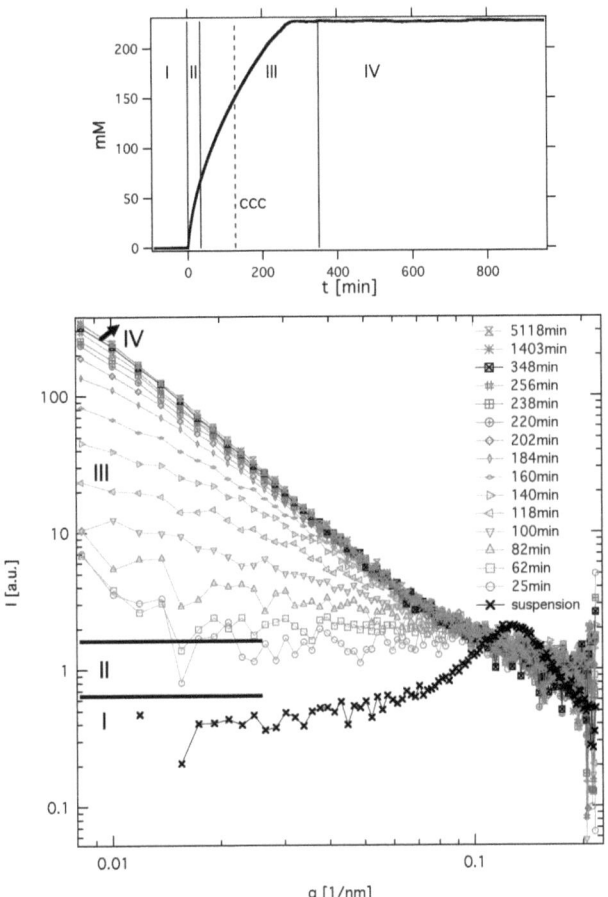

Figure 4.1: Lower graph: raw SANS data of a time resolved gelation process (urea/urease destabilised 19nm PS colloidal suspension with $\phi = 0.038$). Four different phases (I-IV) can be distinguished (see text). In the upper graph, the four phases are superposed with the time dependent ion concentration of the corresponding urea/urease process (graph 2.3). The dotted line marks the critical coagulation concentration ccc.

start of the gelation process shows this characteristics and is similar to the form factor of the particle as the suspension can be considered as an effective hard sphere system and is only weakly correlated at this moment.

In a third phase (III), the particle charges are screened to the extent that the DLVO potential barrier is low enough to be overcome by the thermal energy of the particles. This causes the onset of particle aggregation i.e. the formation and growth of clusters, first following the RLCA and further on (when the potential barrier vanished) the DLCA mechanism. The increasing mass of the growing clusters is reflected in the increase of the forward scattering beyond the intensities of the form factor of the particles (see curves 62min to 348min). Because the clusters grow in a fractal manner, the inner structure becomes visible as a power law dependence of the scattered intensity from q (see equation (1.29)) in the corresponding q range. In the double logarithmic diagram, this is easily identified as a linear slope growing in length with time as the fractal part of the structure of the growing cluster extends to larger length scales. Each of these files can be fitted with the Fisher-Burford structure factor and therefore the radius of gyration of the clusters can be extracted for each curve.

The fourth and last phase (IV) is the gel phase just after the sol-gel transition, where the structure freezes in because the cluster interconnect to a volume filling network. This point is reached at around 350 minutes, and subsequent measurements (shown here are only 1403 min and 5118 min) show no changes in the structure.

4.2 Dynamics

The same sample as measured above with SANS was simultaneously investigated dynamically using light scattering. The DLS data presented in figure 4.2 was taken during the first phases of the aggregation and gelation process, when the sample is transparent enough to assume that single light scattering methods are applicable. The first file shows the suspension before urease

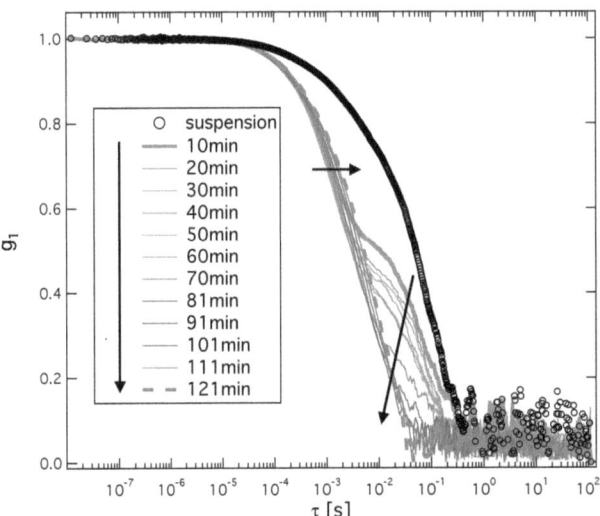

Figure 4.2: DLS data of the same sample as in figure 4.1

was added, representing the highly correlated liquid referred to as phase one above (a detailed discussion follows in section 5.1). The intensity correlation function consists of two superposed decays which are better distinguishable in the second phase (where the screening sets in). At 10 minutes after the start of the ion production, the second decay features a much lower amplitude due to the weaker correlation of the liquid. With rising ion concentration, the amplitude of the second decay decreases but is still quite pronounced when the sample enters its third phase and starts to aggregate at around 60 minutes. The further measurements show two effects: the second decay vanishes completely at approximately 90 minutes, and the cluster formation and growth is reflected by the first decay which shifts slightly to higher lag times. Although the cluster growth should be reflected in far larger shifts towards high lag times, the data can be explained by the fact that the apparent hydrodynamic radius R_h of repulsive particles decreases with increasing volume fraction ϕ. As the effective volume fraction ϕ_{eff} grows during the aggregation process, the resulting decrease of the apparent R_h compensates to a larger extent the effective increase of the cluster size.

The growing clusters increase the turbidity of the sample to a point, where DLS is not applicable anymore. The light scattering part of the combined setup is then switched into DWS mode; the measured data is shown in figure 4.3. The upper graph a) contains the normalised raw data, where the first three correlation functions (142, 157 and 172 min) decay from 1 to zero, indicating that the system is still ergodic. Taking a close look at the first part of the decay, the cluster growth is visible in the shift of the decay to higher lag times. At 187 minutes, the sample clearly exhibits non ergodic behavior by the onset of a plateau. The rotational second cell device forces the correlation functions of the non ergodic sample in a slow second decay at high lag times to zero in order to measure the correct plateau height. The plateau height increases with time and indicates that the dynamics of the sample gets more and more arrested. The correlation function of only the rotational second cell is measured separately by replacing the sample with a

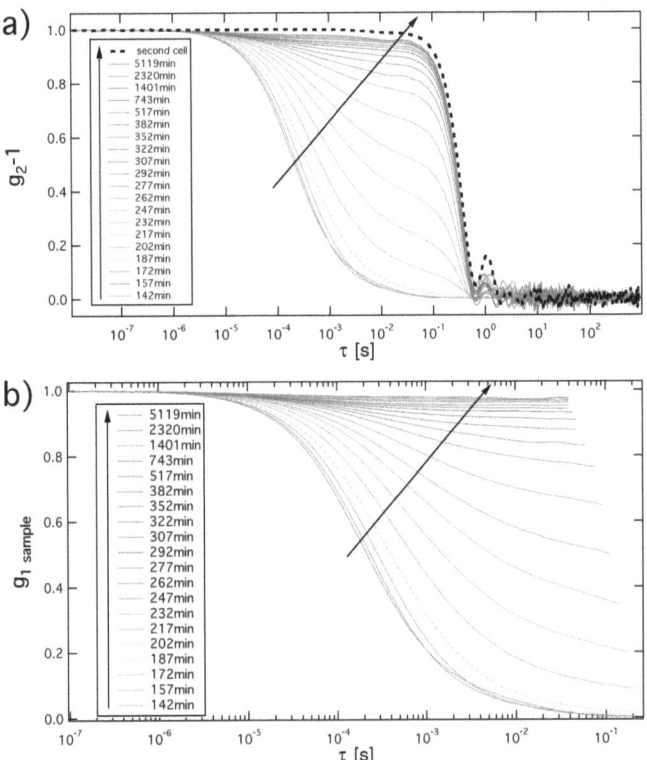

Figure 4.3: a) raw DWS data of the same sample as in figures 4.1 and 4.2, exhibiting two decays: the first decay and the developing plateau are features of the sample, the second decay is caused by the rotational second cell. The correlation function of the second cell alone is measured separately and plotted as a black line for comparison. b) shows the ensemble averaged field correlation function of only the sample which is calculated by dividing the data of a) by the second cell decay. Nicely visible is the onset of a plateau which is the hallmark of restricted motion.

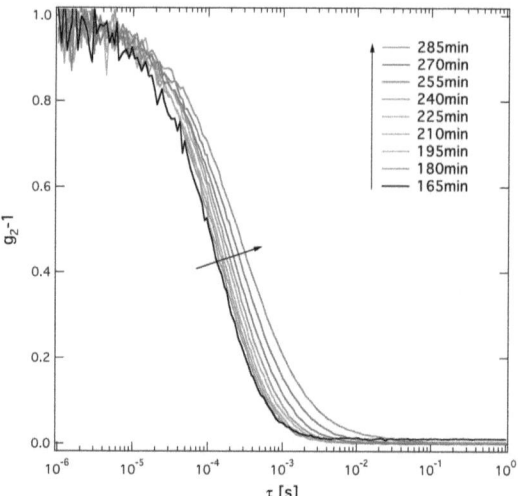

Figure 4.4: DWS raw data of another sample with slower dynamics. The files show the sample in its ergodic state just before the onset of a plateau at the time of 300 minutes. The cluster growth is clearly visible through the shift of the decay to higher lag times.

static scatterer and is plotted in a) for comparison.

Dividing the raw data with the rotational second cell decay, the ensemble averaged field auto correlation function of the sample only can be calculated in a simple way. The resulting $g_{1sample}$ are shown in b).

The DWS data of another sample is shown in figure 4.4 for a better illustration of the cluster growth. The sample is identical to the one above but features a slower aggregation and gelation kinetics (the sample starts to show non ergodic behavior at 300 minutes), allowing us to follow the cluster growth with better resolution in time. The graph shows the evolution of the DWS auto correlation functions in the time range between 165 and 285 minutes. At 165 minutes, the sample gets turbid enough to apply the DWS technique. With time, the decay of the correlation functions moves clearly visibly to higher lag times τ, indicating the growth of the hydrodynamic radius R_H of the clusters. All files up to 285 minutes decay completely from one to zero, indicating that the suspension is still ergodic.

Chapter 5

Discussion

5.1 Highly correlated initial suspension

In its stable initial state, the colloidal suspension chosen for our experiments exhibits a pronounced structure factor in the SANS data (see figure 4.1) and a second decay in the DLS data (see figure 4.2). These observations indicate that the relatively high particle charge results in a strong repulsion and therefore a high positional correlation of the particles. In order to better understand the initial and intermediate state of the sample, tests with varying ion concentrations have been carried out using SANS and DLS.

Starting with the measurement of the initial suspension with SANS (under the conditions described in subsection 3.3.1), the same structure factor peak as already observed before is clearly visible (see figure 5.1 a), sample volume fraction $\phi = 0.038$). With increasing screening of the particle charge through the addition of 1, 5 and 10 mM ions of urea (which is already decomposed by urease), the peak in S(q) gets less pronounced. At the additional ion concentration of 10 mM, the scattered intensity resembles the form factor of the particle at high q and shows the characteristics of only weakly interacting particles. The intensity at low q is still considerably suppressed, but the peak has almost completely disappeared. Also shown in figure 5.1 a) is the form factor of the PS particles as broken line. This form factor is a fit of

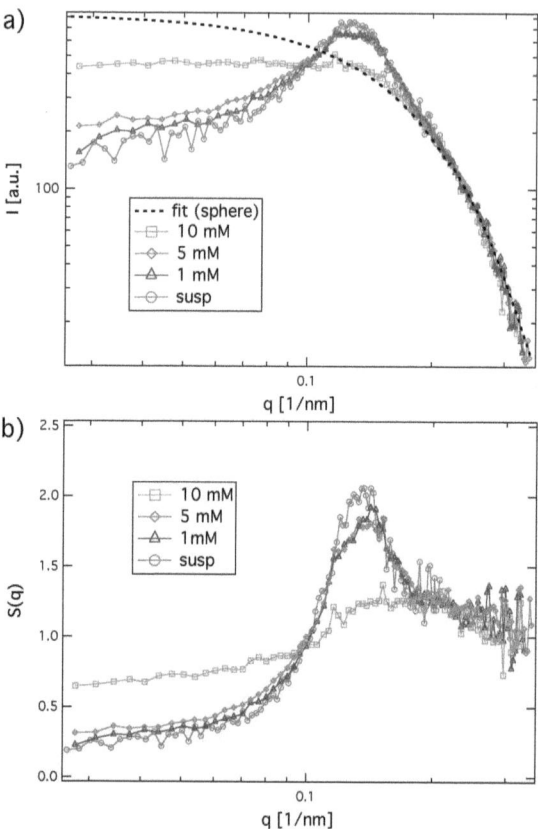

Figure 5.1: a) SANS measurement of the stable PS suspension at different concentrations of added ions (pure suspension, 1, 5 and 10 mM) and the fitted form factor of a dilute sample. The structure peak is clearly visible. By adding decomposed urea/urease to it, the structure peak vanishes. b) The structure factors $S(q)$ calculated by dividing the data of figure a) with the form factor.

data acquired from a dilute solution ($\phi = 0.001$) and takes the particle size polydispersity and apparatus resolution into account. Dividing the scattered intensities from figure a) by the form factor results in the corresponding effective structure factors $S^M(q)$, shown in figure b). The pronounced peak is a typical feature of strongly interacting colloidal particles; it corresponds to a peak in the radial distribution function $g(r)$ (which can be calculated from $S(q)$ through equation (1.28)) and represents the nearest-neighbour shell around a particle (see figure 1.8 C). The peak position $q^* = 0.136 nm^{-1}$ of the initial suspension corresponds approximately to a interparticle distance of 46 nm which is in good agreement with the theoretical estimate using the formulas in 1.2.1. $S(q)$ drops below 1 for $q < q^*$ because of the particle repulsion; particles closer than the first shell are repelled by the Coulomb force. The same argument holds for the shown q range at $q > q^*$ because the particles of the first shell repel the ones of the second shell. Further maxima and minima with decreasing amplitudes corresponding to correlations due to higher order shells are expected to be present, but they all lie outside the shown q range.

We furthermore used the sample containing 10mM decomposed Urea to test whether the strong correlation of the sample in its initial state does influence the gelation process or the static properties of the gel. The strong initial correlation is mostly compensated by the added 10 mM decomposed Urea, but aggregation has still not set in by then (the decomposed Urea has been added 24 hours prior to the measurements and no change in the sample was visible). The same amount of Urea and Urease as used in the standard gelation experiment was added to the sample and the time resolved SANS data then compared to the data of a gelation process starting with the strongly correlated initial suspension. Neither the kinetics nor the structure of the gelled sample showed measurable differences.

Another set of samples with a volume fraction of $\phi = 0.038$ was investigated in round quartz cuvettes of 5mm diameter using a commercial light scattering system (ALV/DLS/SLS-5000F single mode fiber compact

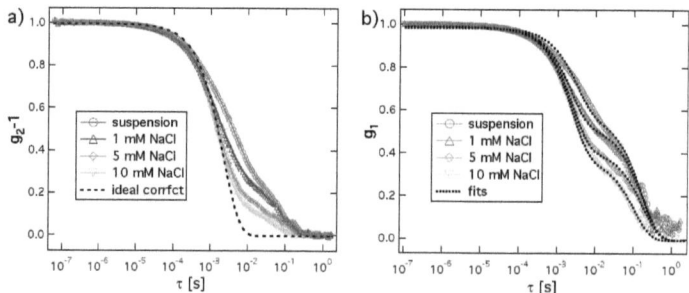

Figure 5.2: a) DLS measurement of the stable PS suspension at $\Theta = 15°$. A second decay is clearly visible. By adding NaCl in concentrations from 1 to 10 mM to it, the second decay diminishes. The calculated ideal correlation function for particles with a diameter of 19nm fits nicely the first decay of the measurements. b) The field correlation functions $g_1(\tau)$ of the data in figure a) and the fitted theoretical functions according to Pusey et al. (see text for details).

goniometer system). At low scattering angles, the normalised intensity auto correlation function exhibits a two step decay (see figure 5.2 a)). The height of the second decay decreases with increasing ion concentration caused by the addition of 1 to 10 mM NaCl. The calculated ideal correlation function for non-interacting spherical particles of 19nm diameter is plotted for comparison of the time scales as a dotted line and corresponds to the first decay of the measured curves. The second decay is a slower process which vanishes when the interparticle correlation gets weaker i.e. the structure factor vanishes. Such a double-decay in DLS can be caused by strongly interacting colloids which feature a size polydispersity. The first decay still corresponds to the usual translative collective diffusion where the density fluctuations at relatively large length scales are probed. The additional second decay may be caused by local relative fluctuations which can only relax through self diffusion of the individual particles. This resembles very much the incoherent part in neutron scattering and can be regarded as a incoherent contribution for light scattering [33]. The self diffusion is slower than the collective diffusion and can only be measured if the individual particles are not identical. Particle size polydispersity and the resulting distribution of scattering power results in the discernability of the particles and consequently in the appearance of self diffusion in the correlation function. Besides the discernibility of the particles, two other conditions have to be fulfilled in order to be able to observe a second decay. First, as the coefficients of both collective and self diffusion are close to each other if the particle interaction is weak (resp. equal when there is no interaction at all), a second decay is only observable for polydisperse systems with relatively strong particle interaction. Moreover the relative amplitudes of the two contributions are given by the polydispersity and the structure factor $S(0)$. Therefore only strong interactions combined with a high polydispersity result in a pronounced second decay.

Pusey et al. [34], [35] calculated the correlation functions including the polydispersity σ of the particles and found for the simplest approximation, the so-called scattering power polydispersity model

$$g_1(q,\tau) = (1-\chi)S^I(0)\exp\left(-D_c q^2 \tau\right) + \chi \exp\left(-D_s q^2 \tau\right) \quad (5.1)$$

where D_c stands for the usual translative collective diffusion coefficient (due to long wave fluctuations in the total number density) and D_s for the self diffusion coefficient (caused by fluctuations in the local relative concentrations of the particles that decay via self diffusion). The parameter χ is defined as [33]

$$\chi = 1 - \frac{(\overline{a^3})^2}{\overline{a^6}} \quad (5.2)$$

where a is the particle radius and $^-$ stands for the arithmetic mean value. The arithmetic value is calculated through $\overline{v} = \int f(v) v \, dv$ where $f(v)$ is the probability density function of the variable v. Assuming a Gaussian distribution of the particle radius, the equation above can then be rewritten

as
$$\chi = 1 - \frac{\left(\int \frac{1}{\sqrt{2\pi}\sigma} \exp\left(-\frac{(a-\mu)^2}{2\sigma^2}\right) a^3 da\right)^2}{\int \frac{1}{\sqrt{2\pi}\sigma} \exp\left(-\frac{(a-\mu)^2}{2\sigma^2}\right) a^6 da} \tag{5.3}$$

With an average radius of $\mu = 9.5$nm and an absolute polydispersity of $\sigma = 1.55$nm (given by the manufacturer IDC), the numerical integration of both integrals in the range of 0 to 50nm results in $\chi = 0.185$. $S^I(0)$ is the ideal structure factor at $q = 0$ and can be calculated in the polydisperse case as
$$S^I(0) = \frac{S^M(0) - \chi}{1 - \chi} \tag{5.4}$$
using the measured structure factor $S^M(0)$ which is calculated as follows
$$S^M(q) = \frac{I^M(q)/\phi}{I^M_{ff}(q)/\phi_{ff}} \tag{5.5}$$

$I^M(q)$ stands for the measured scattered intensity of the sample and the subscript $_{ff}$ for the form factor measurement (separate measurement of the same sample but highly diluted). Figure 5.2 b) shows the field correlation functions $g_1(\tau)$ of the data in figure a) and the corresponding fitted functions according to Pusey et al. (equation (5.1)) where χ is held as a fixed parameter with the value calculated above and $S^I(0)$, D_c and D_s are fitted as free parameters. Using equation (5.4), $S^M(0)_{DLSfit}$ is then calculated from $S^I(0)$ and compared to the $S^M(0)_{SANS}$ extracted from the measured SANS data in figure 5.3. Although a difference of nearly a factor 2 is found at 0, 1 and 5 mM, the dependence on the ion concentration is qualitatively the same: the more ions are added to the sample, the lower the intercept of the second decay is in the DLS measurements and the higher the measured $S^M(0)$ is in the SANS data. The data points at 10 mM also follow this trend, but are slightly off (specially the one of the SANS measurement). We believe that the discrepancies between both measurements arise from the fact that different samples and different kinds of ions has been used. Nevertheless, the experimental data seems to confirm the predictions of the model of Pusey et al. in a consistent and plausible way. Furthermore, the fitted D_c from the

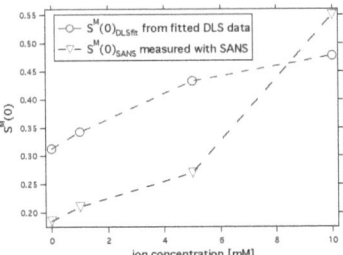

Figure 5.3: Comparison of $S^M(0)$ from the fitted DLS data in figure 5.2 b) and the measured SANS data in figure 5.1 b).

DLS measurements correspond well to the expected value for the particles used and D_s is a factor of 50 smaller than D_c. Pusey et al. [34] reported a factor 3 between their diffusion coefficients, but used larger particles of 50nm diameter with $\sigma=0.19$ and a charge of 1.2 $\mu C/cm^2$ (details given in [36]) at a volume fraction of $\phi=0.0012$. As our sample features particles with 50% higher surface charge density at a 32 times higher volume fraction, the particle interactions should be comparably much stronger and our result appears to be reasonable. If the second decay would be an effect caused by a second population of particles, their radius would correspond roughly to 400 nm and their scattering intensity would have to diminish somehow with increasing ion concentration of the solvent which is hard to imagine.

Figure 5.4 demonstrates that the second decay is caused by a diffusive process. Measurements of a sample without added salt demonstrates that this second decay obeys a q^2-dependence. The slope of the second decay in a log/lin representation of $g_2(\tau) - 1$ versus τ remains constant, indicating a q-independent diffusion coefficient at these characteristic length scales.

Having had so far a look at phase I of the gelation process (see figure 4.1), where the stable suspension was found to be highly correlated and where the intensity correlation functions exhibits a double decay most probably due to high particle repulsion combined with a large particle size distribution, we

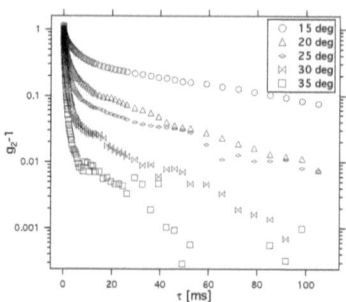

Figure 5.4: Log/lin-plot of the DLS correlation functions of the initial salt-free 19nm PS suspension. The second decay exhibits for different angles the same slope i.e. has the same self diffusion coefficient D_s.

will now proceed to phases II and III where the stable suspension looses its correlation and aggregation occurs because of the addition of ion. We will start with the evaluation of the SANS data which provides information about the cluster radius R_c and the fractal dimension d_f. The approach chosen for our data set is a scaling method, described in the following chapter.

5.2 S(q) scaling method

The self-similarity of the cluster structure allows us to evaluate the time resolved SANS data using a scaling method. The corresponding parts of the structure can be scaled on top of each other by multiplication of the files with shift factors I_{shift} and q_{shift} as follows

$$I' = I \cdot I_{shift} \quad (5.6)$$
$$q' = q \cdot q_{shift} \quad (5.7)$$

This results in a master curve as illustrated in figure 5.5. The upper graph a) shows the same data as figure 4.1 but also includes the fitted particle form factor of a dilute sample of the same particles (thick red line). The

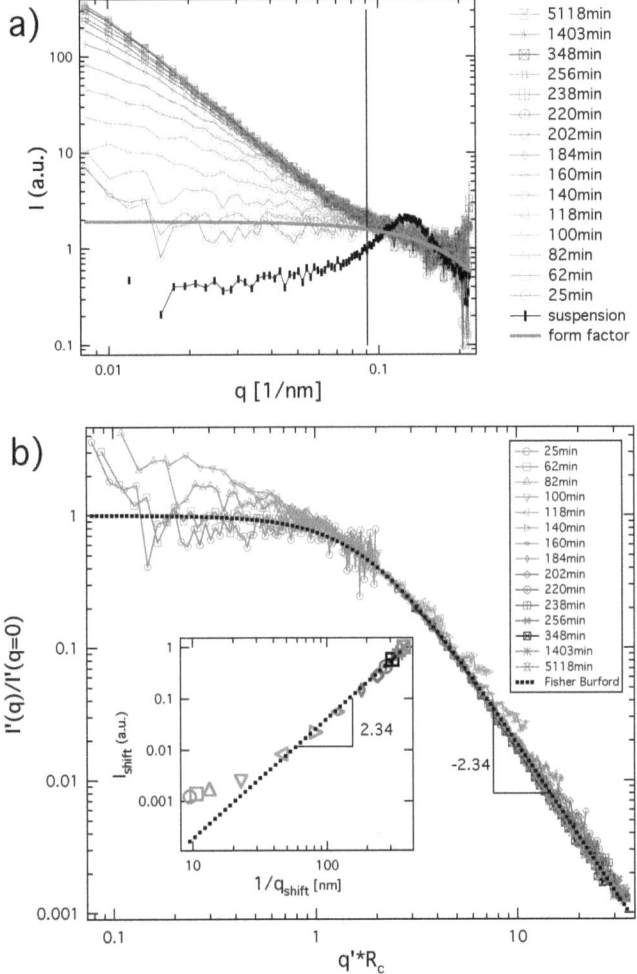

Figure 5.5: Shifted data from the experiment shown in figure 4.1. All files are shifted on top of each other by multiplication. Inset: shift factors.

black vertical line indicates two separate regions: on the left at low q, the universal fractal cluster behavior is observed in the double logarithmic plot in form of a linear slope which stretches with increasing cluster size. Another effect of the growing size of the clusters is the fact that the Guinier regime moves to lower q. The particle form factor of the monomer is constant in this region. On the right side of the vertical line, the probed length scale is below the minimal size for a cluster to exhibit fractal structure, hence local correlations and the crossover to the particle form factor are probed. Being mainly interested in the fractal aspects of the sample, only the left part of the data set was used and shifted on top of each other (see lower graph b)). Starting with the last file, all precedent files are consequently scaled for best overlap of fractal slope and bend-over to the Guinier regime. The introduced scaling factors are interpreted as follows: I_{shift} shifts the measured files to the same scattering intensity, the one of the master curve. As the scattered intensity at low q is proportional to the cluster mass m_c and the measured system is self similar (i.e. their curves have the same shape for the q-range where the self similar structures are measured), the shift in intensity of the measured data to the master curve is proportional to the cluster mass i.e.

$$m_c \propto I_{shift} \qquad (5.8)$$

q_{shift} on the other hand shifts the bend-over of the files to the bend-over of the master file in q direction. As the fractal regime bends over to the Guinier regime at the q-range which is inversely proportional to the corresponding cluster radius R_c, the cluster radius therefore is proportional to q_{shift}^{-1} and hence

$$R_c \propto \frac{1}{q_{shift}} \qquad (5.9)$$

The Fisher-Burford structure factor (formula (1.30)) nicely fits the scaled data set (dotted line). For the lower graph b), the master curve has been scaled to the normalised Fisher-Burford structure factor. The rather large exponent of 2.34 of the slope corresponds to the fractal dimension d_f. It is noticeably higher than the theoretical predictions for DLCA (1.8) and

RLCA (2.1) but in accordance with experimental results of other systems [37, 38, 39]. Furthermore, recent computer simulations of fractal aggregates from concentrated colloidal suspensions [40] show for irreversible aggregation a crossover from DLCA behavior with d_f=1.75 to a percolation regime with d_f=2.5 depending on the interpenetration of the clusters. Whereas our results cannot be explained by pure DLCA or percolation mechanisms, it could very well be interpreted as an intermediate behaviour (see also [41]).

A good way to estimate if the scaling method can reasonably be applied is to investigate the shift factors I_{shift} and q_{shift}. Their values are not independent but have to follow the same power law like the raw data. As seen in the inset of graph b) in figure 5.5, the shift factors clearly show this dependency at large $1/q_{shift}$ i.e. where the clusters are large enough to exhibit an internal structure.

The static data acquired using SANS has shown the decrease of correlation, an onset of destabilisation and cluster growth in agreement with the fractal aggregation model, and the "freezing in" of the structure at long times. The initial suspension and early stage of aggregation could be monitored with DLS, but the sample soon becomes too turbid, and we have to switch to DWS. However, a quantitative analysis of DWS requires additional input, which we now look at in the next chapter.

5.3 DWS analysis

The analysis of our DWS data is not straight forward. In the following subsections, we will first tackle the question whether the DWS method can be applied at all to systems with colloidal particles of such a small diameter as used in this study. Then we will explain how the transport mean free path l^* is estimated, before the applicability of the DWS method for systems at the lower limit of the multiple scattering regime (i.e. at low l^*) is discussed. Having considered these points, we will then calculate the mean square displacement $\langle \Delta r^2 \rangle$ from the DWS data and analyse it with an according model.

5.3.1 DWS applied on small particles

DWS works best in the limit of highly turbid systems where light is scattered from individual particles with a size comparable or larger than the wavelength, $k_0 a \geqslant 1$. For $k_0 a > 1$, DWS measures the (average) single particle mean square displacement while for smaller sizes contributions from collective modes become increasingly important [13]. No attempt has been made however, to our knowledge, to analyze the dynamics of systems consisting of nanosized particles, i.e. where $k_0 a \ll 1$. In a stable suspension such systems appear almost transparent at all densities and are not accessible to DWS. However the large clusters and colloidal gels result in an increased turbidity and hence in principle it should be possible to obtain valuable information about the local cluster and gel dynamics from DWS. In the following we will discuss dynamic multiple scattering of light from fractal aggregates and derive an expression for DWS from colloidal gels.

In a seminal paper Krall and Weitz have developed a simple model to describe the internal dynamics of fractal colloidal gels. They derive the following expression for the intermediate scattering function [5, 21]:

$$F(q,\tau) = S(q) e^{-q^2 \delta^2 [-\exp[-(\tau/\tau_\delta)^p]]} \qquad (5.10)$$

In their model $\delta^2 (1 - e^{(\tau/\tau_\delta)^p})$ is identified as the mean square displacement $\langle \Delta r^2(\tau) \rangle$ of a subcluster (equation (1.63)) of typical size $\xi \cong 1/q$ as long as $1/q \gg a$ ($p \approx 0.7$). This means that scattering of visible light probes the collective motion of particles in the cluster on this length scale[9]. On the other hand if the size of the individual particles is comparable to the wavelength (hence $1/q < a$) both DWS and DLS will probe the mean square displacement of the individual monomer particles in the gel (provided that the inter scatterer distance l^* is sufficiently large compared to any dynamic correlation length in the gel) [17, 18].

---

[9]Since DWS detects q-averaged fluctuations one should note that the DWS-signal predominantly comprises contributions from $q \approx k_0 \to 2k_0$ corresponding to scattering angles $\theta = 60° \to 180°$ due to the q^3 weighing in the averaging integral (see equation (5.14)).

We now want to generalise the application of DWS to the case of (nano-sized) colloidal gels. We start the derivation of the DWS field-autocorrelation function by considering an individual multiple scattering path with N scattering events (analogous to the method described in [13]). In the case of multiple scattering of light the contribution to the normalized field autocorrelation function of a path of length $s = Nl$ is given by

$$g_1^N(\tau) = \left[\int_0^{2k_0} qP(q)S(q,t)\mathrm{d}q\right]^N / \left[\int_0^{2k_0} qP(q)S(q)\mathrm{d}q\right]^N \tag{5.11}$$

with the normalised field-autocorrelation function given by

$$g_1(\tau) \propto \sum_{N=1}^{\infty} P(N) g_1^N(\tau) \tag{5.12}$$

Due to the typically large number of scattering events N along a multiple scattering path, already small values of $[1 - S(q,\tau)/S(q)]$ will lead to a full decay of the DWS correlation function $g_1^N(\tau)$, hence $\delta^2(1 - e^{(\tau/\tau_\delta)^p}) \ll 1$ and therefore

$$S(q,\tau) \cong S(q)\left[1 - q^2\delta^2\left(1 - e^{-(\tau/\tau_\delta)^p}\right)\right] \tag{5.13}$$

hence

$$g_1^N(\tau) = \left[1 - \frac{\int_0^{2k_0} q^3 P(q)S(q)\left[\delta^2\left(1 - e^{-(\tau/\tau_\delta)^p}\right)\right]\mathrm{d}q}{\int_0^{2k_0} qP(q)S(q)\mathrm{d}q}\right]^N \tag{5.14}$$

or

$$g_1^N(\tau) = \exp\left(-N\delta^2\left(1 - e^{-(\tau/\tau_\delta)^p}\right)\frac{\int_0^{2k_0} q^3 P(q)S(q)\mathrm{d}q}{\int_0^{2k_0} qP(q)S(q)\mathrm{d}q}\right) \tag{5.15}$$

which can be rewritten using the relation

$$\frac{l^*}{l} = \frac{1}{3}k_0^2 \frac{\int_0^{2k_0} qP(q)S(q)\mathrm{d}q}{\int_0^{2k_0} q^3 P(q)S(q)\mathrm{d}q} \tag{5.16}$$

to

$$g_1^N(\tau) = \exp\left(-N\delta^2\left(1 - e^{-(\tau/\tau_\delta)^p}\right)\frac{k_0^2 l}{3l^*}\right) \tag{5.17}$$

with $N = s/l$ we obtain for the field auto-correlation function (replacing the sum by an integral)

$$g_1(\tau) = \int_0^\infty P(s) \exp\left(-\frac{1}{3}k_0^2 \frac{s}{l^*}\delta^2 \left(1 - e^{-(\tau/\tau_\delta)^p}\right)\right) ds \qquad (5.18)$$

Taking relation (1.63) into account, this expression has the same structure as equation (1.56) and leads to the same functional form of the DWS correlation function in transmission geometry (equations (1.58) resp. (1.60)). Therefore the DWS method can be applied to measure (in a fractal gel) the motion of individual particles which are small compared to the wavelength. But in order to do so, the transport mean free path l^* needs to be known. This is not always a trivial task and will be discussed in the following subsection.

5.3.2 Calculation of l^* from SANS data for DWS

For the correct estimation of the mean square displacement $\langle \Delta r^2(\tau) \rangle$ using DWS in transmission geometry, l^* needs to be known. This parameter changes drastically during the gelation process. But it is difficult to measure the direct transmission of light in this setup, because the transmitted light beam lies within the neutron beam and is therefore not easily accessible for detectors. Even if there would be space enough to mount another aluminum coated silica wafer as a mirror, this would disturb the neutron measurements too much, and the ambient light would tamper the measured transmission intensities especially in the important range of high turbidity. Moreover, comparative measurements with reference samples would be cumbersome. However, there is another way to estimate l^* from the measured data: it is possible to calculate l^* from the static data by estimating the change of the structure factor. From theory, l^* is calculated using following general equation:

$$l^* = \frac{k_0^4}{\pi \rho} \left(\int_0^{2k_0} S(q) P(q) \left(1 + \cos^2(\theta)\right) q^3 dq \right)^{-1} \qquad (5.19)$$

Where $S(q)P(q)(1 + \cos^2(\theta))$ stands for the total anisotropically scattered light intensity normalised with the total intensity of the individual particles (see equation (1.26)). The additional term $(1+\cos^2(\theta))$ can usually be neglected as it is equal to 1 if the light is scattered in a scattering plane perpendicular to the polarisation of the light. But the path of a photon through a turbid medium is not restricted to the scattering plane and resembles more a random walk in three dimensions, and therefore the anisotropic nature of light scattering has to be taken into account.

Comparing the l^* of an uncorrelated suspension (l^*_{susp}) with the l^* of a gel (l^*_{gel}) at a given gel age, constants will cancel. Furthermore, the range of integration covers a q range where the Fisher-Burford structure factor describes the scattered intensity sufficiently accurate. In this case, we replace in equation (5.19) the traditionally used expression $S(q)P(q)$ for colloidal suspension with the intensities $I(q)_{gel}$ and $I(q)_{susp}$ for the q dependent scattering intensity of the gel and the suspension, resulting in the following equation

$$\frac{l^*_{gel}}{l^*_{susp}} = \left(\frac{\int_0^{2k_0} I(q)_{gel}\,(1+\cos^2(\theta))\,q^3 \mathrm{d}q}{\int_0^{2k_0} I(q)_{susp}\,(1+\cos^2(\theta))\,q^3 \mathrm{d}q}\right)^{-1} \quad (5.20)$$

The scattered intensity of the suspension $I_{susp} \propto S(q)_{susp}(1+\cos^2(\theta))$ is nearly constant in the range from 0 to $2k_0$ (the particles are very small, so the form factor is not varying at those q values, and the structure factor of an uncorrelated suspension is a constant as well), and therefore the integral yields a constant value when calculated. Using trigonometric relations and equation (1.13) it is possible to reformulate $\cos^2(\theta) = (1-\frac{q^2}{2k_0^2})^2$ and we find:

$$l^*_{gel} \approx l^*_{susp} \frac{\frac{16}{3}k_0^4}{\int_0^{2k_0} \frac{I(q)_{gel}}{I(q)_{susp}}\left(1+(1-\frac{q^2}{2k_0^2})^2\right)q^3 \mathrm{d}q} \quad (5.21)$$

l^*_{susp} is determined using a program based on Mie scattering of polydisperse spheres. With known scattered intensities in the integration range, l^* can be deduced. The ratios between structure and form factors in neutron scattering are equal to the structure and form factor ratios of light scattering for our system because we work in both cases in the Raleigh-Gans-Debye limit ($2kR|n_1/n_2 - 1| \ll 1$). For this reason, we can replace $\frac{I(q)_{gel}}{I(q)_{susp}}$ in equation

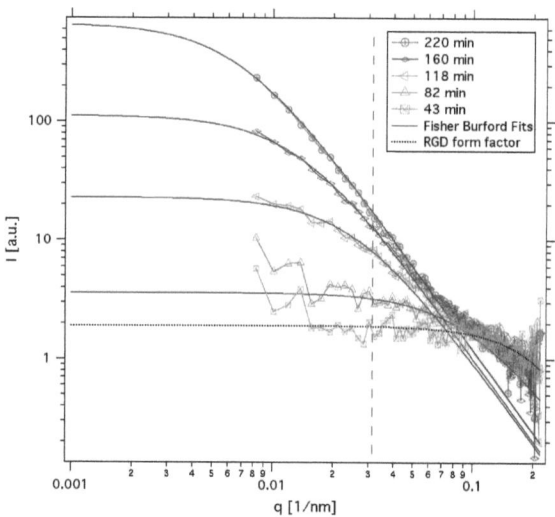

Figure 5.6: Selected SANS data from figure 4.1 and according Fisher/Burford fits for the calculation of l^*. The integration range is $q < 2k_0$ (on the left side of the dashed vertical line), where the form factor of the suspension is nearly constant and therefore can be treated as a constant value.

(5.21) with the corresponding values for neutron scattering. As seen in figure 5.6, the scattered neutron intensity of the uncorrelated suspension (which corresponds to a RGD fit of the file file at $t=43$min where the suspension is decorrelated) is not varying in the integration range $0 < q < 2k_0$. The parameters used for the Fisher-Burford structure factors are estimated using the scaling method described in the section above and checked against the real data. The extrapolation of these structure factors to q=0 provides the ratio of $\frac{I(q)_{gel}}{I(q)_{susp}}$ over the needed q range.

The l^* calculated as described above is shown in figure 5.7. Because the values were found to be relatively high for our samples, we have to estimate the critical condition below which the DWS approach is justified and data

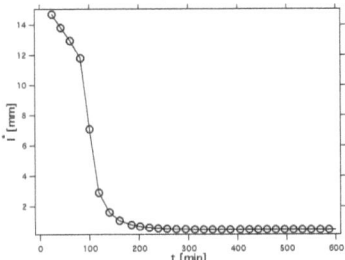

Figure 5.7: l^* calculated from the SANS data presented in figure 4.1.

analysis produces reasonable results. This will be discussed in the following subsection.

5.3.3 DWS applicability at low L/l^*

As mentioned in chapter 1.2.2, DWS theory can only be applied to turbid samples. The rule of thumb is that reasonably quantitative results can be calculated for $(L/l^*) \geqslant 5$. Nevertheless, from a careful analysis information can be obtained for samples starting at $(L/l^*) \approx 2$. To show the applicability of the DWS method at such low L/l^*, a well defined model system was diluted from volume fraction ϕ=0.02 to 0.001. This resulted in a calculated reduction of L/l^* from 15 to 0.77. We deduced from the known particle radius the transport mean free path l^* using equations (1.60), (1.48) and (1.49). The system used was a colloidal suspension of charge stabilised polystyrene spheres with a diameter of 0.78 μm (manufactured at IDC). The measurements were done in transmission geometry with the same detector angle (45°) as the combined SANS/DWS setup, the laser wavelength was λ=532nm.

As seen in figure 5.8, the measured L/l^* starts to deviate from the calculated one at around the point 2.39 L/l^*. Nicely visible is also the strong increase of the measured scattered intensity for the scattering angle of 45° once the suspension starts to exhibit increasing multiple scattering (the multiple scat-

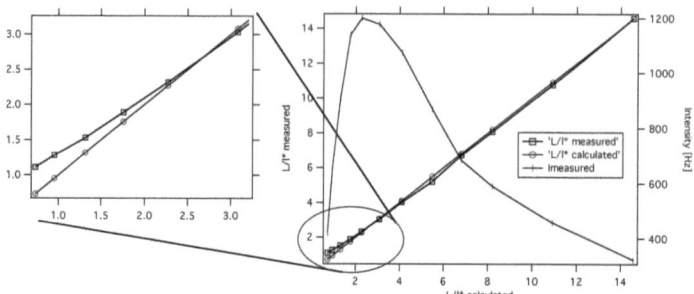

Figure 5.8: Comparison of measured (DWS method) and calculated L/l^* in a model system: a deviation is visible for $L/l^* < 2.39$, as seen in the blow up. This value corresponds to the point where the scattered intensity (measured at a scattering angle of 45°) reaches its maximum. To lower values of L/l^*, the intensity decreases because the incident beam is scattered less by the sample. It is in this regime where the DWS results vary from the calculated values because the number of scattering events is too low to justify the use of the diffusion equation for the photon transport.

tering events increase the number of photons scattered from the direct beam in direction of the detector). At L/l^*=2.39, the intensity reaches its maximum and then decreases because the large number of scattering events leads to increasing reflection from the sample.

Having shown in the precedent subsections that the application of DWS to our system is valid in the range above $(L/l^*) \approx 2$ and having explained how the important time dependent parameter l^* can be deduced from our static measurements, we can now proceed by evaluating the measured DWS data.

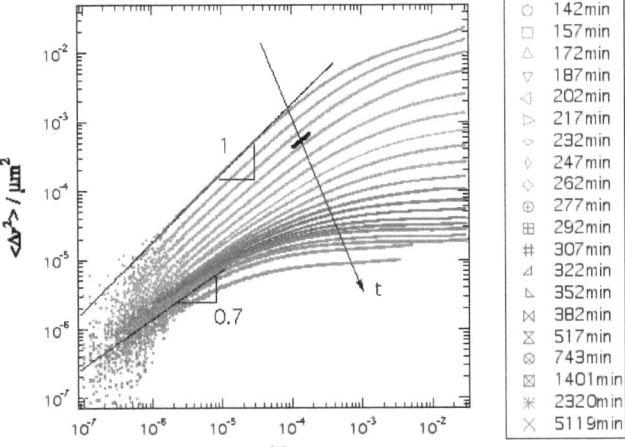

Figure 5.9: Mean square displacement of the correlation data in figure 4.3. The time increases in direction of the arrow; the short line indicates the point where the system starts to exhibit non ergodic behavior.

5.3.4 Mean square displacement $\langle \Delta r^2 \rangle$

To discuss the DWS data shown in figure 4.3, the mean square displacement (MSD) $\langle \Delta r^2 \rangle$ of each file is calculated using an l^* estimated with the method described in subsection 5.3.2 and applying equation (1.60). The non ergodic correlation functions have been divided by the second cell decay prior to the MSD calculation. The time resolved MSD shown in figure 5.9 evolve in time in direction of the arrow.

The cluster growth is visible in the slow-down of the MSD with time: as large clusters move over shorter length scales than small ones during the same time, the MSD for a chosen lag time τ decreases with growing size. The initial slope p of the MSD decreases from 1 to 0.7, indicating a qualitative change of the particle motion from diffusive (with the theoretical value of p=1) to sub-diffusive behavior. Sub-diffusive motion for colloidal gels exhibits a power-law with exponent p=0.7 (Krall and Weitz [5]), in nice agreement with our data. Another feature of this data set is the onset of a plateau for the MSD, a characteristic sign of confined movement. A short line in the arrow designating the temporal evolution marks the point where the onset of a plateau is visible in the DWS raw data. The MSD data doesn't feature such a clear distinctive point but a nearly continuous change from diffusive to constrained movement, therefore the DWS raw data was used to identify the crossover from ergodic to non ergodic behavior. The decrease of the initial slope and the development of a plateau in the MSD data represents particles whose motions are sub-diffuse and restricted to short distances. The RLCA/DLCA theory explains this through the fact that the particles are trapped in a network. With time, the network builds up more connections and gains in rigidity, and the distances which the particles (fixed in the network positions) can explore become shorter, therefore the MSD plateau lowers its height. The particles also start to "feel" the restrictions of the network at shorter time scales, causing the cross-over time τ_c to move to shorter lag times τ.

We applied a scaling method similar to the one described in section 5.2 to the MSD of figure 5.9, starting with the file at 277 minutes. As shown in figure 5.10, they all collapse to a master curve with an initial slope of 0.7 which represens subdiffusive motion. The dotted line plotted in the same figure represents the theoretical curve of the simple model of Krall and Weitz for colloidal gels (equation (1.63)). Although the crossover from subdiffusive to arrested motion found in our data is not as sharp as in the Krall/Weitz model, the qualitative behavior is nicely reproduced.

5.4 Gelation kinetics

The time-resolved raw data of SANS and DWS discussed separately in section 4.1 and subsection 5.3.4 exhibit quite nicely the behavior expected by theory

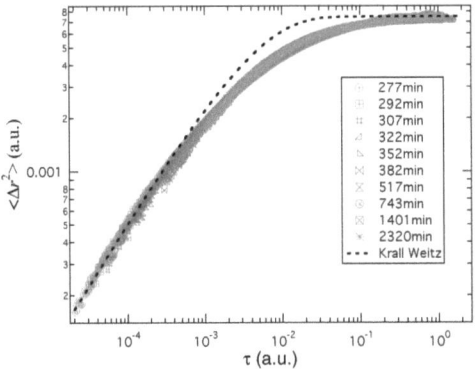

Figure 5.10: Scaled mean square displacement curves of figure 5.9, starting with the file at 277 minutes. The dotted line is a guide to the eye and represents the theoretical curve for a gel according to the Krall/Weitz model.

and show no unknown features. But when the dynamic and static data are combined in the same graph, a very interesting development of the gel kinetics appears. The static cluster radius R_c evaluated using the shifting method on the SANS data and the initial slope p of the MSD are shown in figure 5.11 as a function of the time t ($t = 0$ at the start of the urea/urease process).

The first measured "cluster" radius $R_c = 9.5 nm$ corresponds exactly to the particle radius given by the manufacturer, and the final value of $R_c = 380 nm$ is in reasonable agreement with the estimation of $R_{c,critical}$ for fractal systems (equation (1.8)): using a pre-factor of 0.3 (as found by Weitz et al. [5] for similar conditions), $R_{c,critical} = 404 nm$ for $a = 9.5 nm$, $\phi = 0.038$ and $d_f = 2.34$. Following to theory, the development of R_c before the critical coagulation concentration ccc at 127 minutes should obey an exponential law as the system is in RLCA conditions (equation (1.7)). After ccc, the cluster radius is expected to grow following a power law under DLCA conditions

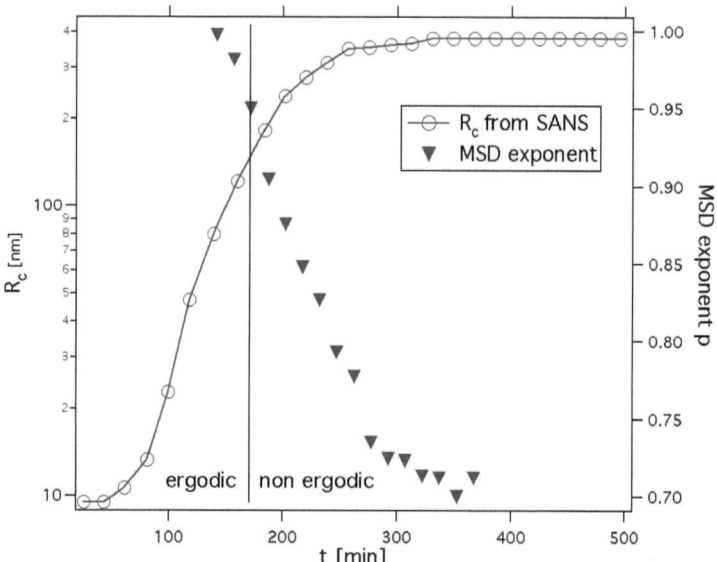

Figure 5.11: Combined data from static and dynamic measurements: the static cluster radii from SANS (circles) and the slope p of the mean square displacement of all DWS files (inverted triangles). The vertical line indicates the point where the system starts to exhibit non ergodic behavior in the dynamic data.

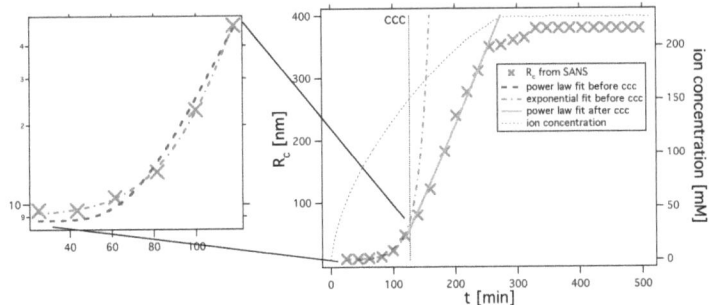

Figure 5.12: R_c from the SANS measurements (crosses) fitted before ccc with an exponential function and power law. As seen in the inset on the left side, the exponential law fits the data nicely. After ccc, a power law fit describes the curve best.

(equation (1.6)).

Shown in figure 5.12 are two fits of R_c below ccc, one for an exponential function and one for a power law. As there are only few data points and the difference between both fits is small, it is difficult to make an unambiguous statement about the form of the cluster growth. But the inset on the left side of the graph shows that the exponential function fits the data most accurately. The evolution of R_c above ccc is fitted with a power law which describes well the cluster growth for the steep ascent up to 254 minutes, but with an exponent of 1.07 it is only slightly different from linear behavior. According to equation (1.6), the growth should be a lot slower with an exponent $1/d_f$ which equals 0.56 in the case of $d_f = 1.8$ (for pure DLCA). It is important to keep in mind that the ion concentration is continuously increasing during the gelation process which does influence the cluster growth rate. Moreover, the scaling analysis of the SANS data indicates a higher fractal dimension of approx. 2.34, which is also compatible with the crossover from DLCA to a percolation regime seen in the computer simulations of irreversible aggregation of comparable volume fractions for $R_c \sim 5a - 10a$ [40]. Percolation would also predict a power law growth of the fractal cluster with an exponent of 1.7 [42]. The experimentally observed growth kinetics is thus between the predictions for DLCA and percolation, respectively.

A very interesting point is the fact that the simultaneously measured DWS data set shows all characteristics of restricted motion after 172 minutes (vertical black line in figure 5.11) i.e. becomes non ergodic at this point. Surprisingly, this is clearly before R_c reaches the plateau with its final value. This means that R_c continues to increase after the system changes from ergodic to non ergodic behavior. This behavior is not congruent with the RLCA/DLCA theory, where the cluster grow until they connect to a network, forming a gel at this point. The formation of a system spanning network structure defines in theory the time at which the system changes from ergodic to non ergodic. But after the formation of such a network, clusters cannot grow anymore; in contrary, if the measured characteristic length scale would change, then it would rather decrease due to the increasing number of connections. Therefore the onset of non-ergodicity cannot be interpreted as the creation of a network in our case, because the clusters still continue to grow afterwards. Taking a look at the measured exponent p of the initial MSD slope, the change from p=1 to p=0.7 parallels the evolution of R_c: first, diffusive cluster aggregation takes place with increasing R_c and p=1, and in the end when R_c stops to grow, the predicted value of p=0.7 for subdiffusive motion of particles in a colloidal gel is reached. But the onset of subdiffusive motion (i.e. $p < 1$) occurs at a time where R_c is still growing (and therefore the motion should be freely diffusive according to theory), in perfect agreement with the transition from ergodic to non ergodic behavior. This behavior can only be understood by the introduction of an intermediate stage.

We thus propose an intermediate state of the system, illustrated in figure 5.13. After the stable colloidal suspension a) and the aggregation stage b) where clusters still undergo free diffusive Brownian motion, a glass like state

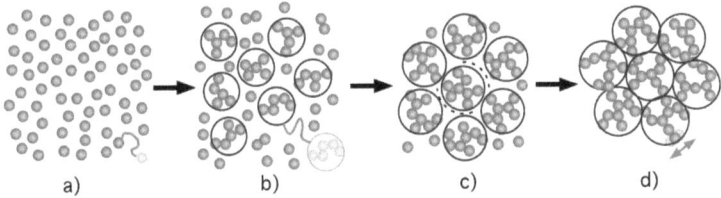

Figure 5.13: Gelation process including a glass like state c) between cluster growth in free diffusion b) and interconnected gel state d).

c) is established before the cluster interconnect and build a network d). In the glass like state, the clusters grew already large enough to hinder each other to move around freely, but they are still not interconnected (this is the so-called cage effect). In this trapped state, a two step decay with a plateau would be expected for the dynamic data, but the cluster would have still room enough to grow. Such a glass like state before the gel point was already proposed by Segré et al. [30] and Bissig [29].

5.4.1 Additional Echo DWS measurements

To learn more about the dynamics of such systems at very long correlation times (i.e. slow processes), the Two-cell Echo approach (see paragraph 1.2.2) was used. The accessible lag times $0.1s < \tau < 450s$ (for our chosen parameters) are much larger than the fast decay of the sample, where in the double cell technique the slow decay of the second cell starts to set in and therefore covers any information of the sample. Using the Two-cell Echo approach, the particularly interesting question whether the correlation function of the sample exhibits a diverging second decay at large τ (as expected a glass transition) or a plateau at constant height (as expected for a gel) can be adressed. But as this technique is limited to relatively long lag times, it can only be used in combination with other methods like for example the double-cell technique for a full description of the sample dynamics.

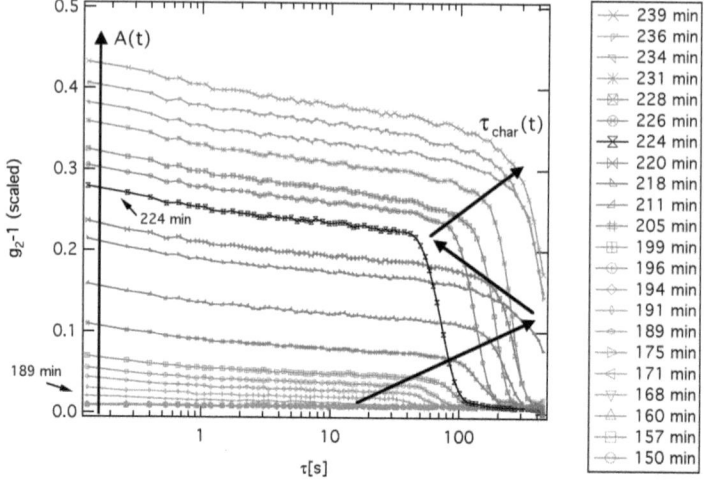

Figure 5.14: Data of the second decay in a time resolved gelation process, measured with a new DWS technique called Two-cell Echo approach. The plateau height A(t) increases continuously with time, while the characteristic time $\tau_{char}(t)$ first increases, then drops drastically backwards and then increases again.

The raw data of a measurement using the Two-cell Echo approach is shown in figure 5.14. The sample is a colloidal suspension of 19nm PS with a volume fraction of $\phi = 0.038$ undergoing a urea/urease induced gelation process with same parameters and under the same conditions as described in section 2.3. The data contains only the second decay, the so-called α decay in the case of glasses. The first decay (β decay) takes place on shorter time scales and is therefore not measurable with this method. To find the correct plateau height A(t), the plateau heights of the echo data have been estimated by scaling the echo plateau height of the gelled echo sample (ca. 24 hours after the process started) to the plateau height of the DWS measurements of the sample from section 4.2 at the same gel age. Figure 5.14 can therefore be regarded as the extension of figure 4.3 b).

Clearly visible is the onset of a plateau at 189 minutes. With time, the plateau height increases continuously. The signal decorrelates with a characteristic time τ_{char}. In a first stage from 189 to 218 minutes, τ_{char} increases monotonically. Then the characteristic time drastically decreases (220 and 224 minutes) but continues to grow afterwards again in a second stage after which it seems to settle around an average value. To investigate this behavior, the gelation time t of the echo files is scaled to the gelation time of the files of the sample from section 4.2 through comparison of their plateau heights. The reproducibility of the gelation process is not perfect (the gelling times may vary up to a factor of 3 in extreme cases) but qualitatively sufficient to justify the scaling approach.

The cluster radius R_c from SANS and the corresponding plateau heights of the simultaneously measured DWS data are shown in figure 5.15 a). The vertical line indicates the time of the onset of a plateau. The plateau height increases strongly during the first 100 minutes after the onset, indicating an increasing nonergodicity or "hardening" of the system. This is mainly due to the increasing effective volume fraction caused by the cluster growth, and because of the increasing attraction between the particles (i.e. the smaller Coulomb repulsion) due to the still rising ion concentration. After the first 100 minutes, the growth of the nonergodicity parameter (the plateau height) slows down. It is then mainly driven by the increasing rigidity of the network through the creation of new bonds. By scaling the plateau heights of the echo measurements to the plateau heights of the SANS/DWS data set, the equivalent gel age t and cluster radius R_c of the echo measurement relative to the SANS/DWS data can be estimated. Since l^* is varying after the onset of the plateau, the correlation functions have been corrected for the sample turbidity by multiplication of the lag time τ with $(L/l^*)^2$. The scaled and turbidity corrected characteristic times τ_{char} of the echo measurement are estimated by shifting the files relative to each other (similar to the scaling

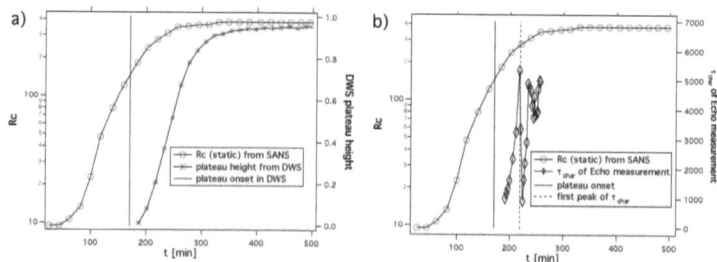

Figure 5.15: Scaling of the echo data with the SANS and DWS data of section 4.1. a) the cluster radius R_c from SANS and the corresponding plateau heights of the simultaneously measured DWS data. The vertical line indicates the time of the onset of a plateau. b) the same SANS data (R_c) and the scaled echo data, showing the characteristic time τ_{char} of the second decay (turbidity corrected).

method described in section 5.2). In figure 5.15 b), τ_{char} is plotted relative to R_c. τ_{char} first seems to diverge and then collapses after a first maximum. The first maximum is indicated by the dotted line and corresponds to a time where the fast cluster growth is finished. The subsequent increase of τ_{char} then settles at a certain height, showing no more abrupt changes. In figure 5.16, τ_{char} is plotted against the effective volume fraction

$$\phi_{eff} = \phi \left(\frac{R_c}{a}\right)^{3-d_f} \tag{5.22}$$

ϕ_{eff} is very sensitive with respect to d_f; varying d_f from 2.34 to 2.2 changes the values of ϕ_{eff} by 35%. Therefore already minor experimental and analytical errors may have a strong influence on the quantitative result. We interpret the initial increase of τ_{char} as due to the increase in effective volume fraction owing to the growth of space filling clusters. Similarly to the hard sphere glass transition we observe an apparent divergence of τ_{char} at some critical volume fraction. Above that apparent critical point we find that the correlation function still decays, and that the characteristic decay time

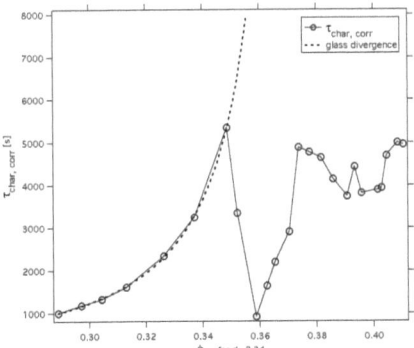

Figure 5.16: The characteristic time τ_{char} of the echo measurement (as already shown in figure 5.15 b)), corrected for turbidity and plotted against the calculated ϕ_{eff}. The dotted line is an exponential fit of the first 7 data points and represents the glass divergence.

first decreases an then increases again. This decay is most likely related to the formation of the first percolating paths, where we can expect that large elastic modes propagate throughout the sample, effectively contributing to dephase the scattered light. The characteristic cut-off of these modes should become smaller as the network gradually becomes more connected, leading to an increase in the characteristic time.

A very successful model for the description of the glass transition is given by the mode coupling theory (MCT) [43, 44] which is based on the nonlinear coupling (feedback mechanism) of the microscopic dynamics i.e. different density fluctuation modes. It predicts for increasing coupling strength a sharp ergodic to non-ergodic transition where the fluid becomes permanently frozen. Both α and β relaxation processes close to and beyond this transition as well as their time scaling properties can be described using MCT, providing an accurate qualitative explanation of the cage effect. MCT is a relatively complex theory and doing quantitative calculations within its framework is quite complicated but doable. The characteristic time τ_α of the α decay follows the power law

$$\tau_\alpha \propto (\phi_c - \phi)^\alpha \qquad (5.23)$$

Where ϕ_c is the critical volume fraction at which τ_α diverges. The glass divergence (dotted line) in figure 5.16 is an actual fit of the data using the equation above, resulting in fitted parameters of $\alpha = 1.54$ and $\phi_c = 0.38$. The parameter α is in good agreement with theory, where a value of $\alpha > 1.5$ is expected [45]. ϕ_c on the other hand cannot be accurately compared to experimental data from literature (given as $\phi_c = 0.56 - 0.58$ for hard spheres [44, 46]) because ϕ_c varies for different samples and therefore the volume fractions of all experimental data is usually rescaled using a relation $\phi_{reduced} = (\phi_c - \phi_{eff})/\phi_c$ as already discussed by Pham et al. [47]. Pham et al. investigated the structure and dynamics of the simplest model colloid, a suspension of hard spheres (PMMA). An attractive force was induced by adding a polymer (linear PS), creating an effective depletion attraction. Among other experiments, echo measurements of samples with ϕ varying around the glass transition concentration were carried out. By changing the amount of added polymer (and therefore changing the strength of the attractive force), the change from a repulsive glass to a liquid and furthermore to an attractive glass at the same particle volume fraction could be observed (the so-called re-entrant behavior). As our system shows a divergence of τ_α between 189 and 218 minutes (corresponding to 189 resp. 205 mM, which is clearly above ccc$\simeq$150 mM), the colloidal particles bear a short range attraction (see figure 2.1) and therefore form an attractive glass. The phase diagram of Pham et al. (figure 1 in [47]) shows in addition to the re-entrant behavior the existence of attractive glasses at high polymer concentrations (i.e. strong attractions) for volume fractions down to ϕ=0.36. In view of the considerable uncertainty our measured ϕ_{eff}, its fitted value of 0.38 seems therefore to be in reasonable agreement with literature.

In short, the Two-cell Echo approach allows us to measure the correlation function at long lag times, showing the plateau and the α decay of our sample.

After the onset of a plateau in the correlation function, τ_{char} grows in a first stage according to a power law. We interpret this as glass like dynamics, in reasonable accordance with MCT theory. This behavior is shown in a period of time, where the DWS data of the first decay exhibit a plateau with increasing height, but the SANS data still reflects substantial growth. A glassy phase would explain this behavior as the clusters are constrained in motion by the surrounding clusters but still have a possibility to grow through a percolation like process, where bonds between clusters are created. The fact that the α and β decay are quite far apart is probably caused by the attraction of the particles, linking the cage of the surrounding clusters stronger together and lowering the probability of an "escape".

Chapter 6

Finite gels with varying Φ

So far, we discussed only gels with volume fractions of ϕ=0.038. We also performed time-resolved neutron and light scattering on samples at different ϕ but found them mostly unsuitable for our aim for the following reasons: at both high and low volume fractions, DWS is not applicable anymore because the samples are not turbid enough. In the case of low ϕ the number of particles is too low, and for high volume fractions the cluster radii R_c are too small to scatter light strongly enough. With respect to time-resolved SANS, the measurement time of 15 minutes per file is not sufficiently long to provide data with good statistics for samples with low ϕ, and samples with high volume fractions gel too fast. Because we were aiming to measure samples time-resolved and simultaneously with both techniques in order to get information about structure and dynamics, we restricted our study to the volume fraction of ϕ=0.038 (as discussed in the whole work up to now).

Nevertheless, particle gels of different volume fractions were characterised with SANS. We measured a series of colloidal gels in the range of $0.01 < \phi < 0.18$ in order to calculate their structure factors. These measurements were undertaken in view of previous studies that reported a breakdown of the "fractal" description of the DWS results for samples with volume fractions above 10%. Romer [48] was the first to observe this effect for colloidal gels by measuring the dynamics of the gelation process with DWS, and Bissig [29] continued her work. He investigated destabilised colloidal systems (PS with a=85nm, destabilised with Urea/Urease) in the range of $0.01 < \phi < 0.25$ using DWS and analysed it using equation (1.63). He found that the dependence of τ_δ and δ^2 on ϕ exhibits deviations from the power law behavior expected for fractal gels at $\phi \sim 0.10$. Poulin et al. [49] also reported a deviation from the predicted behavior at high volume fractions. Their experiment consisted of SLS studies on destabilised concentrated emulsions which showed at low ϕ the expected cluster growth but no further change of cluster size above $\phi \sim 0.10$.

In order to test our system for this anomaly, the structure of colloidal gels was measured using SANS. Samples with various volume fractions have been prepared by dilution and concentration of the 19nm PS suspension described in chapter 2. The solvent of all samples is identical to the one of the gelling samples of the previous chapter (52% H_2O and 48% D_2O). The rather delicate process of concentrating the samples was carried out using a ultrafiltration unit (Amicon Stirred Cell Model 8010 with polyethersulfone membrane PBVK02510). The 19nm PS stock solution(with pure H_2O as solvent) was first diluted with D_2O and then filled into the ultrafiltration unit. Inert gas (nitrogen) with a pressure of ca. 3 bar pressed the solvent through a membrane which held back the colloidal particles. The volume fraction of the resulting concentrated suspension was estimated by weighing a part of the suspension and comparing it to the weight of the same part of the suspension after evaporation of the solvent. Tests showed no difference in the structure of gels which have been destabilised with different amounts of urease (from 20 up to 200 u/ml). This allowed us to choose faster ion production rates to speed up the gelation process. The structure factors $S(q)$ calculated as described in section 5.1 are shown in figure 6.1. The q range for which the gel exhibits fractal behavior is easily identified as it is represented in a log-log plot through a linear shape. At low volume fractions, where the individual cluster is free to grow over long distances in a fractal manner before it gets restricted by the presence of other clusters, the fractal

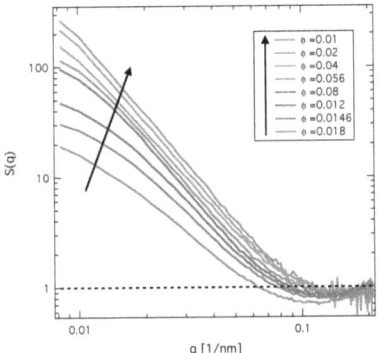

Figure 6.1: The structure factor of finite gels with varying volume fraction ϕ.

range spreads at least over a decade of q. Because at higher ϕ the growing clusters will hinder each other at shorter length scales, the fractal range gets shorter (see also the discussion on page 19) and the turnover to the guinier regime becomes visible at low q, indicating the growing cluster size. The "correlation hole" at the right end of the fractal regime where the structure factor drops below 1 gets deeper with increasing ϕ as expected. The data shown in figure 6.1 demonstrates nicely the continuous decrease of the fractal range (and the continuous cluster growth) as well as the deepening of the correlation hole as predicted by theory, showing no signs of an anomaly.

Furthermore, we analysed the structure factors of figure 6.1 using the scaling technique described in section 5.2. Figure 6.2 shows the scaled structure factors which are fitted with the Fisher/Burfod structure factor. We remark that the fractal dimension d_f equals -2.38 and is very close to -2.34 which we found for the time-resolved gelation process of a sample at ϕ=0.038. d_f is therefore considerably higher than predicted in traditional theories such as DLCA but still below the fractal dimension of percolating systems (see discussion on page 87). From the shift factors and the Fisher/Burford fit,

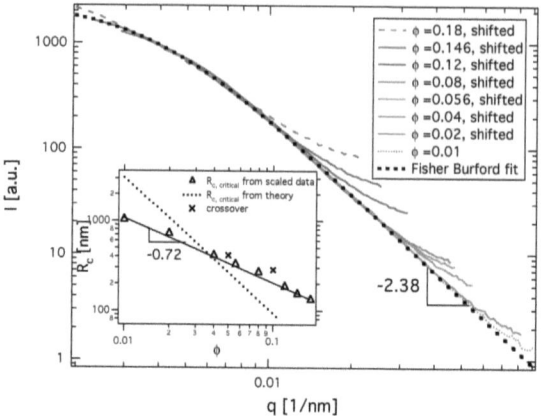

Figure 6.2: Shifted structure factors of finite gels with varying volume fraction ϕ. The inset shows the resulting cluster radii $R_{c,cluster}$.

$R_{c,critical}$ can be calculated. The shifted data file for ϕ=0.18 shows a slight deviation from the Fisher/Burford fit at low q values and was therefore additionally analysed with a Guinier plot. The resulting $R_{c,critical}$ is only 10% smaller compared to the one from the scaling technique. The inset of figure 6.2 shows the critical cluster radii of the scaled data as well as theoretical values calculated as described on page 97. The fitted cluster radii seem to follow a power law with an exponent of -0.72 which would corresponds to a fractal dimension of d_f=1.6. We thus find a clear deviation between the slope of the experimental data and the theoretical slope given by 1/(3 - df) = 0.83 for DLCA and 2 for percolation. It is clear that the experimental structure factors at particle volume fractions below 0.04 show no clear turnover from the fractal to the Guinier regime in our measured q range. Therefore the shift factors cannot be evaluated precisely and our fitted values for the critical cluster radii $R_{c,critical}$ thus bear a large error. Another uncertainty arises from the multiplication factor of 0.3 which was used for the calculation as estimated by Weitz et al. based on measurements with colloidal systems at low volume fractions. However, while this results in an additional uncertainty for the absolute values of $R_{c,critical}$, it will influence the calculated curve $R_{c,critical}$ vs. ϕ in a double logarithmic representation via a shift factor only, but will not have any influence on the slope, provided that the prefactor is indeed concentration independent over the entire concentration range.

It is interesting to compare our values for the critical cluster radius $R_{c,critical}$ at different volume fractions with recent Monte Carlo computer simulations of irreversible (and reversible) aggregation at comparable volume fractions [40]. These simulations demonstrate the existence of a maximum cluster radius based on a crossover from percolation clusters (d_f=2.5) to a homogeneous system (d_f=3). The maximum cluster radius shows a dependence on volume fraction in reasonable agreement with our data (shown as crosses in the inset). However, it is clear that we lack enough data of sufficient accuracy in the low volume fraction to reach a conclusive statement. It would certainly be interesting to perform additional light and small-angle light scattering experiments at low volume fractions to investigate the crossover from a percolation regime to the DLCA regime known to exist at very low volume fractions. Nevertheless, we have been able to unambiguously demonstrate a monotonic decrease of $R_{c,critical}$ for systems up to volume fractions of ϕ=0.18. For our system we therefore cannot confirm a cessation of the cluster growth at a volume fraction of 0.1 as observed for emulsions by Poulin et al., and the structure of our gels yields no indication of a structural anomaly or transition which could explain the observations of the dynamics in particle gels made by Bissig.

Chapter 7

Conclusion

In this work, we investigated a model system to gain further insight into the gelation process of systems with high volume fractions. A colloidal suspension of small latex particles was therefore destabilised using the so-called Urea/Urease process. The controlled and homogeneous in situ production of ions allowed us to prolong the gelation process to a few hours, offering us the possibility to follow the gelation kinetics in a series of time resolved measurements. The symbiotic combination of two non-invasive scattering techniques (neutron and light scattering) resulted in the simultaneous acquisition of static and dynamic information of the same sample at similar length scales. Whereas the static neutron scattering can be applied to measure the samples in all phases of the gelation process, the light scattering methods have to be adapted due to the strongly changing optical properties of the sample. In a first stage, the colloidal suspension is nearly transparent due to the small particle size, and dynamic light scattering (DLS) can be applied. After the point where the aggregating clusters reach a certain size and the system becomes sufficiently turbid, diffusing wave spectroscopy (DWS) can be used for further investigations. The standard DWS scheme had to be extended with an improved double cell technique for the correct ensemble averaging when the sample becomes non ergodic, and with a novel DWS echo technique that gives access to very slow processes.

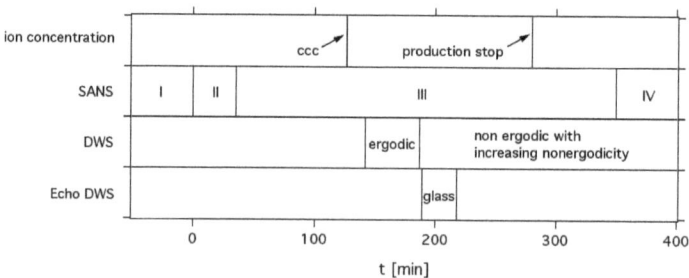

Figure 7.1: Overview of different temporal hallmarks of the gelation process measured with SANS, double cell DWS and two-cell Echo DWS.

An overview of the most important results is shown in figure 7.1. The static data acquired with SANS was split into four phases: phase I describes the stable suspension, where SANS as well as DLS show a strong correlation of our system which can be qualitatively explained by theory. Phase II contains the uncorrelated suspension which lacks a strong structure factor (as expected). In phase III, aggregation and cluster growth occur and finally the structure of the sample stops to evolve i.e. the system is found to be fully gelled in phase IV. At first sight, the sample behavior appears to follow the theoretical predictions from classical aggregation models such as DLCA. The cluster growth in phase III is also in good accordance with theory as the radii from the static measurements initially increase exponentially (like predicted for RLCA) up to the critical coagulation concentration (ccc), where we observe a crossover to a power law growth as predicted for DLCA. However, the simultaneous measurements with double cell DWS revealed a transition from ergodic to non ergodic behavior at an astonishingly early stage, in the middle of phase III where fast cluster growth is still visible. Clearly, this transition cannot be attributed to the formation of a system spanning network of interpenetrating clusters because in this case the characteristic length scale would not have the possibility to grow. We interpret this dynamical

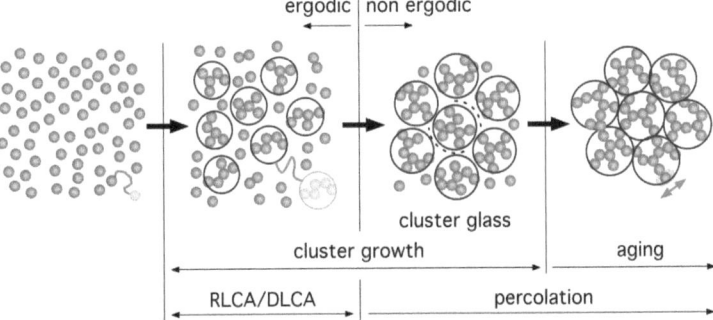

Figure 7.2: Schematic gelation process including a cluster glass phase before the gel.

arrest with the formation of a fractal cluster glass, that then further evolves through percolation-like formation of bonds into a final gel.

If this early ergodic/non ergodic transition would be the beginning of a glassy phase, a diverging α decay at long correlation times should be measurable with DWS. However, the double cell technique introduces a forced second decay of the correlation function that prevents access to any dynamic process that relaxes more slowly. Therefore we had to reinvestigate this regime using a newly developed method, the two-cell Echo DWS. This technique enabled us to measure our system in the relevant time scales for the α decay. The kinetics of the characteristic decay time of the α process strongly resembles a glass divergence and occurs during the last part of the fast cluster growth measured with SANS. We therefore believe that our interpretation is justified and that the gelation process of our system includes a glassy phase after DLCA cluster growth and before the formation of the gel network as shown in figure 7.2.

In conclusion, the combination of SANS and DWS has demonstrated that there exists an intermediate regime in the sol-gel transition inconsistent with the simple DLCA model, where the sample becomes non ergodic, but where the average cluster size continues to grow until it reaches the final value of the maximum cluster size in the gel. The DWS echo measurements have allowed us to identify this intermediate regime as an arrested cluster glass, where the clusters are only able to perform local motion due to cage formation, but where the average cluster size continues to grow through a percolation-like process. This percolation-like process is responsible for the formation of the final bonds which lead to the gel formation and also provides an explanation for the high fractal dimension d_f, in agreement with recent computer simulations of irreversible aggregation and gel formation of comparable volume fractions.

Part II

Multi3D

A new setup was built applying the so-called 3D scattering technique [50, 51]. This technique provides an efficient suppression of multiple scattering by performing simultaneously two independent light scattering experiments in the same sample volume and at the same scattering vector. By cross correlating both signals, only singly scattered light is taken into account for the correlation function. Using this scheme, the application of dynamic and static light scattering (DLS and SLS) can be extended to turbid systems. Being interested in the kinetics of colloidal suspensions undergoing aggregation, gelation and phase separation, the acquisition time should be as small as possible, but certainly smaller than the characteristic time scale on which the system changes its properties. Measuring simultaneously at four different angles reduces the acquisition time by a factor of four, and DLS measurements can be enhanced in precision by taking into account the weighing of size distributions at the different measured angles [52, 53]. Considering these advantages, we have constructed a multi angle 3D cross correlation setup (Multi3D) to investigate dense nano- and mesostructured complex fluids and solids.

Chapter 8

Setup

8.1 Theory 3D DLS and SLS

The well established DLS and SLS techniques (see subsection 1.2.2) provide valuable dynamic and structural information about a sample, but are based on single light scattering and therefore limited in their application. Systems exhibiting single scattering appear to be transparent, a rare condition for a large number of the samples colloid scientists would like to investigate. To avoid multiple scattering, different tricks like index matching, sample dilution and scattering path length reduction have been used. This is of course not unproblematic as the sample characteristics may be altered. Therefore, different approaches to suppress multiple scattering based on cross correlation have been developed in the last years [51].

The cross correlation function $G_{12}(\tau)$ of the intensities I_1 and I_2 can be written as

$$G_{12}(\tau) = \langle I_1(t) I_2(t+\tau) \rangle \tag{8.1}$$

If we simultaneously perform two scattering experiments (with two initial beams with wave vectors $\vec{k_{01}}$ and $\vec{k_{02}}$ and two detectors at final wave vectors $\vec{k_{s1}}$ and $\vec{k_{s2}}$) in the same scattering volume with identical scattering vector $\vec{q}$ but different geometry, we obtain the two intensities I_1 and I_2. We then find the following relation between the auto correlation function $g_1(q,\tau)$, the measured intensity auto correlation function $G_{11}^{(1)}(\tau)$ and the measured intensity cross correlation function $G_{12}^{(1)}(\tau)$ in the absence of multiple scattering

$$G_{11}^{(1)}(\tau) = I_1^{(1)2}\left(1 + \beta_{11}|g_1(q,\tau)|^2\right) \quad (8.2)$$

$$G_{12}^{(1)}(\tau) = I_1^{(1)}I_2^{(1)}\left(1 + \beta_{12}|g_1(q,\tau)|^2\right) \quad (8.3)$$

where $^{(1)}$ denotes singly scattered photons. The intercept β_{11} of the autocorrelation function depends mainly on the detection optics whereas the cross correlation intercept β_{12} is furthermore reduced due to phase mismatch $\delta q = |\vec{q}|$ (mismatch between $\vec{q_1}$ and $\vec{q_2}$) and misalignment $\delta x = |\vec{x}|$ (spatial mismatch between both scattering volumes).

$$\beta_{12} = \beta_{11} e^{-\frac{\delta q^2 R'^2}{4}} e^{-\frac{\delta x^2}{R'^2}} \quad (8.4)$$

R' denotes the distance between the scattering volume and the detector. The information contained in the auto and cross correlation function is the same in the case of single scattering. But when multiple scattering occurs, the auto correlation function will include the contribution of multiply scattered photons which makes the deduction of $g_1(q,\tau)$ from $G_{11}(\tau)$ very difficult or even impossible. In the cross correlation experiment on the other hand, only singly scattered light produces correlated intensity fluctuations on both detectors - multiply scattered light results in uncorrelated fluctuations that contributes to the background only due to the fact that it has been scattered in a succession of different q-vectors. The contributions from multiple scattering to the signal are suppressed by a factor of order $(R'\delta k_j)^{-1}$, where δk_j denotes the magnitude of the smallest of the two wave vector combinations $\vec{k_{02}} - \vec{k_{01}}$ and $\vec{k_{02}} + \vec{k_{s1}}$ [51]. The auto correlation function $g_1(q,\tau)$ therefore is related to $G_{12}(\tau)$ through

$$G_{12}(\tau) \approx I_1 I_2 + \beta_{12} I_1^{(1)} I_2^{(1)} |g_1(q,\tau)|^2 \quad (8.5)$$

where I_j is the average total intensity (singly and multiply scattered) measured at detector j and $I_j^{(1)}$ is the singly scattered intensity only. Multiple scattering will therefore only decrease the intercept, but the auto correlation

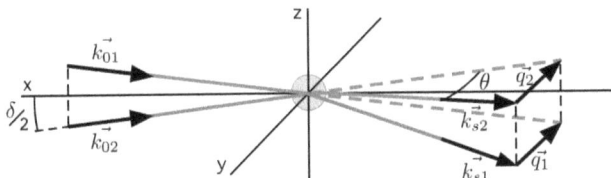

Figure 8.1: The 3D cross correlation scheme: two incident beams with wave vectors $\vec{k_{01}}$ and $\vec{k_{02}}$ are out of the scattering plane xy by an angle of $\delta/2$ (but they lay in plane xz). They cross the scattering plane at exactly the same volume, the scattering volume (symbolised by the grey sphere). The light scattered at the scattering volume by the angle θ is also out of the scattering plane, resulting in wave vectors $\vec{k_{s2}}$ resp. $\vec{k_{s1}}$. Both scattering vectors $\vec{q_2}$ resp. $\vec{q_1}$ are identical and parallel to the scattering plane.

function $g_1(q, \tau)$ can still be measured even for turbid samples using a cross correlation scheme.

Only two of the proposed and successfully implemented scattering geometries for the cross correlation scheme allow for the variation of the scattering angle [54]: the two-colour technique (TCDLS) [55] and the 3-D technique (3DDLS) [56, 57, 58]. In TCDLS, two beams of different wavelength are scattered and subsequently measured using interference filters. The two independent scattering experiments can thus be discerned very well, but the alignment is very difficult (special emphasis should be placed on the fact that for different wavelengths the same $\vec{q}$ is measured only for different scattering angles θ, and this difference $\Delta\theta$ varies with q).

The 3DDLS technique uses another approach to discern the two scattering experiments: the incoming beams and the detectors are tilted by an angle of $\delta/2$ with respect to the usual scattering plane, crossing it in the scattering volume (see figure 8.1). Although the same wavelength is used and therefore each detector measures the scattered intensity of both incoming beams, the scattering vectors of the photons from both incoming beams are different and the beams can be discerned that way. The additional contribution from the "wrong" experiment increases the uncorrelated background i.e. reduces the intercept β, but the advantage of simple and fast scattering angle variation makes 3DDLS the method of choice.

8.1.1 Intercept β and overlap volume

The intercept β can be regarded as the signal to noise ratio of the dynamic light scattering experiment. While the theoretical maximum of $\beta_{11,max}$ for auto correlation experiments is 1, the theoretical maximum of $\beta_{12,max}$ for the 3DDLS technique is 0.25 due to the fact that both incoming beams use the same wavelength and are therefore measured with both detectors. The cross correlation function can be written as

$$G_{12}(\tau) = \langle I_1{}^i(0) I_2{}^i(\tau) \rangle + \langle I_1{}^i(0) I_2{}^{ii}(\tau) \rangle + \\ + \langle I_1{}^{ii}(0) I_2{}^i(\tau) \rangle + \langle I_1{}^{ii}(0) I_2{}^{ii}(\tau) \rangle \quad (8.6)$$

where 1 and 2 denote the detectors and i and ii the incident beams. Only one term - the second one - of the four corresponds to the desired combination of incoming beams and detectors following the 3DDLS scheme. The intensity of the other three terms is still detected, but they produce different scattering vectors $\vec{q_1} \neq \vec{q_2}$ and are thus uncorrelated.

The theoretically obtainable 3DDLS intercept $\beta_{12,max} = 0.25$ can be reduced due to misalignment of the instrument and overlap volume effects. The misalignment (spatial mismatch of the scattering volumes δx and of the scattering vectors δq) lowers β_{12} in the following way

$$\beta_{12} = \beta_{12,max} e^{-\frac{\delta x}{R'}} e^{-\frac{\delta q^2 R'^2}{4}} \quad (8.7)$$

In addition to misalignment, the overlap volume effect has to be taken into account. Figure 8.2 illustrates the overlap volume (i.e. the common volume of both incident laser beams, marked with thick lines) and the scattering volume (i.e. the common volume of the incident beams and the detector beams). As all beams are tilted with respect to the usual scattering plane by

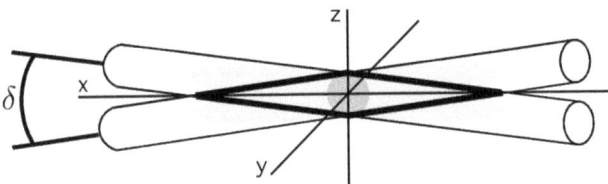

Figure 8.2: The scattering volume (light grey for a small resp. large θ, dark grey for $\theta = 90°$) depends on the scattering angle. The overlap volume (fat lines) is the volume illuminated by both incident beams.

the angle $\delta/2$, the scattering volume will depend on the scattering angle. At the scattering angle $\theta = 90°$, the dark shaded spherical region in figure 8.2 represents the scattering volume (if the detector beams are of the same size as the incident beams). The scattering volume is fully contained in the overlap volume. For all other angles, the scattering volume is in first approximation a factor of $1/\sin(\theta)$ larger (for a small and big scattering angle θ marked as light grey area). A considerably big part of the scattering volume where only one incident beam illuminates the sample is outside the common overlap volume, and therefore the intercept β_{12} decreases. Due to the overlap volume effect, β_{12} has a maximum at $\theta = 90°$ and decreases continuously for larger and smaller scattering angles. This effect can be experimentally determined and has been compared with the corresponding theoretical calculations [59].

8.1.2 SLS on turbid media

The intercept β_{12} described above can be measured using a highly dilute sample in which only single light scattering occurs. The resulting $\beta_{12}^{(1)}$ represents the maximal intercept achievable with the current alignment and is a standard way to calibrate a 3D instrument. In that case, δx and δq are experimentally determined and overlap volume effects are taken into account,

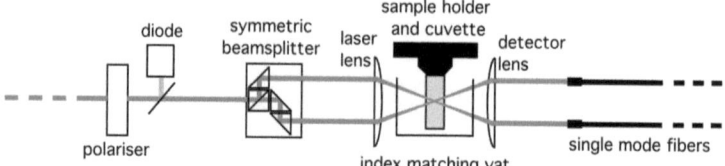

Figure 8.3: Diagram of a 3D cross correlation setup in side view: the incoming beam (from left) passes a vertical polariser and its intensity is measured with a diode. It is then split into two parallel beams which are focussed by a lens into the scattering volume. The scattered photons pass another, symmetrically positioned lens and are then collimated into single mode fibers. The sample is contained in a cuvette which is placed in a index matching vat to reduce refractions on the cuvette and for thermal stabilisation.

therefore any further reduction of the measured intercept β_{12} is caused only by multiple scattering. It is thus possible to distinguish between the intensity caused by single scattering events and the contribution of multiple scattered light through a comparison of β_{12} and $\beta_{12}^{(1)}$. The singly scattered intensity (denoted with the superscript $^{(1)}$) is calculated as follows:

$$I^{(1)}(q) = \sqrt{I_1^{(1)}(q) I_2^{(1)}(q)} = \sqrt{\frac{\beta_{12}}{\beta_{12}^{(1)}} I_1(q) I_2(q)} \qquad (8.8)$$

Due to the distinction between singly and multiply scattered light (which is a special feature of DLS measurements using a cross correlation setup), SLS measurements of turbid samples are possible by performing for each measured scattering angle a short DLS measurement.

8.2 Description setup

The setup was designed and built in-house. Four detector stations are mounted pairwise on two removable platforms on both sides of the incident beam (see figure 8.4 and also picture 8.5). This enables one to align the two detectors on one platform while the other one is removed. Dowel pins assure the precise repositioning of the two platforms. While the platforms have fixed positions, the incident beam can be moved using a goniometer arm.

A HeNe laser provides a beam of the wavelength of $\lambda = 632.8$ nm. It is coupled into a polarisation preserving single mode fiber (Dantec DAN 60x30) whose end is mounted on a goniometer arm (Newport goniometer 496 with Newport Motion Controller PMC400). The beam passes a polariser and glass platelet, where a small fraction of the beam is reflected on a diode (UDT 10DP 9645-1) to measure its intensity. Afterwards the beam is split into two beams; a special arrangement of one beam splitter and three prisms splits the incident beam symmetrically in vertical direction into two beams which run parallel to the direct beam with a distance of 12.5 mm, both beams with an intensity of around 45% of the incident beam. The beams pass the laser lens (1) and are focussed into the sample. The sample is contained in a round cuvette which is positioned in a index matching vat (Hellma, quartz glass, round with outer diameter of 85 mm, custom produced). The index matching fluid used is decalin and it is thermally controlled through a heat exchanger coil connected to a thermostat. Both vat and cuvette positions can be adjusted via separate tilting and x/y translation mechanisms.

The scattered light is focussed on each of the four detector stations by identical detector lenses 2-5 via the built-in collimation lenses into two single mode fibers (OZ LPC-01-532-3.5) which are connected to photomultipliers (PM, Hamamatsu H3460-54 photon counting heads). As high precision must be achieved, all lenses (1-5) are placed in 5 axis holders (Newport M-LP-2) for easy and reproducible tilting, translation and focussing. The Dantec fiber end where the laser beam emerges is also placed in such a holder, and each of the eight detection fibers has a separate tilting and translating holder (a combination of Standa 7T128 x/y translation stages and Newport U50-A tilting mounts). The signals of the PMs are processed by two amplifiers/discriminators (Hamamatsu C3866 photon counting units) and then

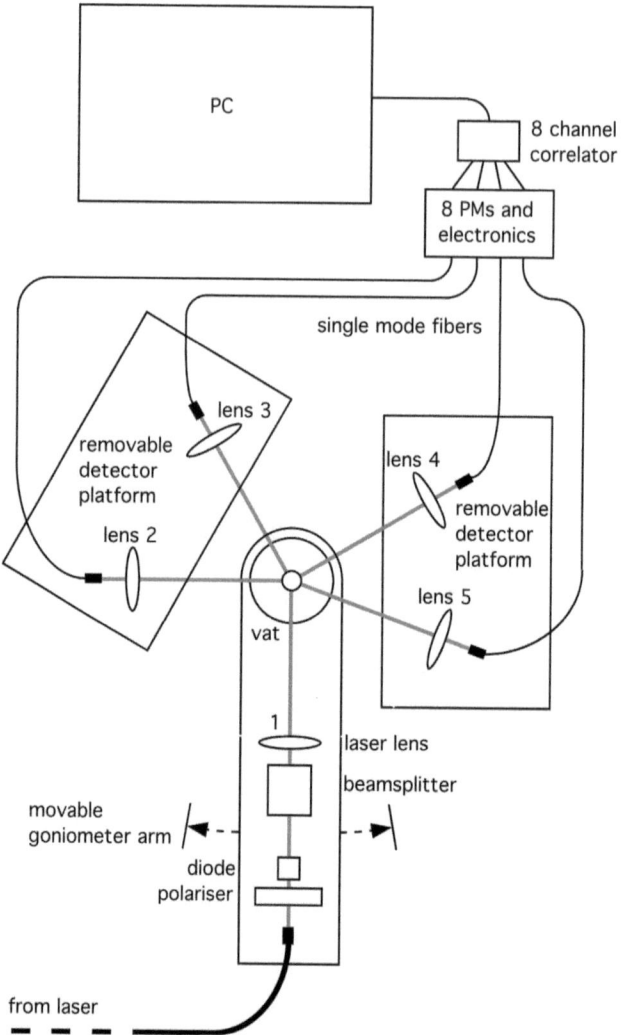

Figure 8.4: Schematic diagram of the Multi3D setup in top view. See text for details.

Figure 8.5: The Multi3D instrument: 4 detector stations at different, fixed angles (1-4), index matching vat and sample holder (5), incoming laser beam, mounted on a goniometer arm (6)

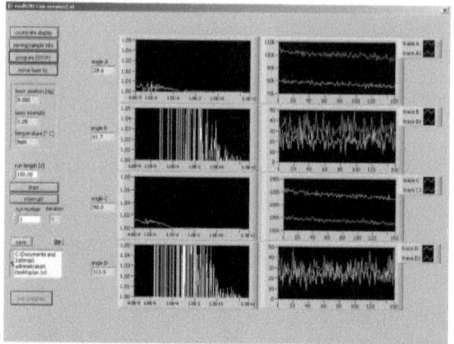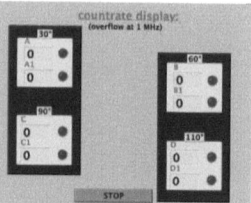

Figure 8.6: Screen shots of the Multi3D software: main window on the left, count rate display on the right.

fed into a digital correlator (correlator.com Flex01/8ch). This correlator is an eight channel correlator which calculates four cross correlation functions in parallel using a multiple tau layout with a minimal lag time of 40 ns. A PC runs a Labview program which reads out the digital correlator through a USB port. The Labview program also controls the goniometer motion controller via RS-232 and a I/O-PCI-card (National Instruments NI 4350 with CB-68T Terminal Block) to read out the laser beam intensity from the diode, the vat temperature trough an immersed Pt100 sensor and the limit switches for the goniometer movement.

8.3 Software

The software is written in Labview 6.1. Its main tasks are to read out, display and save the correlator data and to control the goniometer arm; a screen shot of the main window is shown in figure 8.6 on the left side. A special intensity count window (count display, shown in the same figure on the right side) helps to adjust the scattering intensities; too low values result in bad data statistics and too high intensities can irreversibly damage the photomultipliers. Limit switches stop the goniometer movement to avoid possible misalignment, and temperature and incident laser intensity are measured to allow a correct data evaluation. The program is divided into different modules which are then combined in a envelope. This flexible structure allows for easy and quick modifications or extensions.

8.4 Alignment

The alignment of the instrument is a crucial point and a very demanding task. The necessary steps described in the following subsections are summarised in the schematic figure 8.7. As the beam diameter is around 100 μm and will be focussed into the sample, high precision is necessary to finally be able to measure two different signals with exactly the same scattering vector. Any misalignment leads to loss of intercept of the cross correlation function.

8.4.1 Mechanical alignment

First of all, the mechanical components have to be aligned. The index matching vat is aligned by positioning a micrometer on the goniometer arm and turning it around the vat, measuring and reducing any variation of the relative height (horizontal alignment) and distance of the inner vat wall (centering). Both detector platforms have to be removed to maximise the range of measurement. The sample holder is aligned the same way, but a special centering tool which is basically a round plate on a rod (see figure 8.8) is inserted in place of the sample to allow the micrometer to measure height and the relative elongation from the goniometer axis. The precision of horizontal alignment and centering of vat and sample holder has to be below 10 μm.

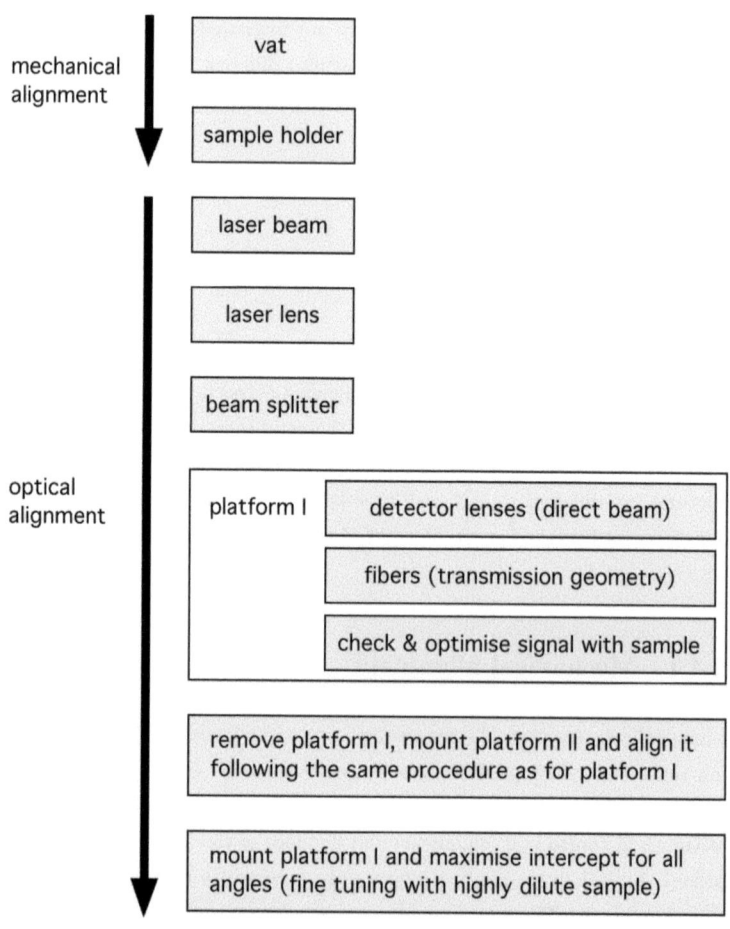

Figure 8.7: Schematic overview of the Multi3D alignment process.

Figure 8.8: Alignment tools for the Multi3D apparatus: left a micrometer with magnetic foot, in the middle a 100 μm pinhole fitted in a holder of the dimensions of a standard cuvette (upside down). On the right, upside down as well, is the special centering tool for the sample holder.

8.4.2 Optical alignment

The incident laser beam has to be aligned first. The vat has to be filled already with the index matching fluid (decalin) to include all refractions which will be present later on. The laser lens (1) and the beamsplitter have to be removed, but the polariser should stay mounted for the same reasons as mentioned before. By adjusting the back-reflection of the incident beam from the vat (for all possible angles), the laser can be aligned normal to the vat surface. A pinhole (diameter 100 μm) fixed in the sample holder is used to check the position of the beam (again at all possible angles). Marking the position of the incident beam, it is possible to insert the laser lens (1) and adjust it in a way that the incident beam, which is supposed to be in the middle of the lens, hits the mark of the beam without lens. After this rough alignment, fine-tuning is done with the pinhole. All three beams (both parallel beams as well as the direct beam, without beamsplitter) have to pass through the pinhole. A symmetric refraction pattern has to be achieved, and stability and symmetric behavior for focussing or x/y translation of the laser lens (1) must be obtained. The correct and proper alignment of the laser lens (1) is vital.

To align the detector lenses (2-5), the goniometer arm is moved to the position in which the beams should directly hit the fibers (transmission geometry). Before doing so, the detector fibers have to be disconnected from the detectors to avoid possible damage. One detector platform has to be mounted on its dowel pins. The pinhole has to be removed from the sample holder and the beamsplitter from the goniometer arm. The incident beam is used as the reference to align the detector lens by marking the beam position without detector lens and adjusting the beam with inserted lens to the same position. After mounting the beamsplitter again, upper and lower beam have to be coupled in to the respective fiber. Doing so for both detector stations on the mounted platform, one half of the instrument is roughly aligned. The alignment of the other platform follows the same procedure, and once both platforms are aligned and mounted, fine adjustment can be started.

The rough optical alignment described above should be precise enough to generate a visible intercept in the cross correlation function when a sample is measured. If this is not the case, the rough alignment has to be repeated. Measuring once a correlation function with an intercept, the latter is increased by measuring a highly diluted sample which guarantees single scattering. Adjusting the detector lens and the fiber positions will result in higher intercept if carefully done. The fibers are extremely sensitive to tilting, and the translation in horizontal direction changes the scattering angles i.e. the intercept varies strongly with this parameter. Cyclic repetitions of this optimisation process should result in intercepts of typically between 0.15 and 0.18 for a scattering angle of $90°$.

Chapter 9

Test measurement

After successful assembly of the setup and thorough testing of all components and the software, the instrument was roughly aligned for testing purposes to half of the maximum possible intercept on two detector stations on the same detector platform. After removal of the aligned platform, another detector station on the other platform was roughly aligned. The repositioning of the first, aligned platform was not very precise but good enough to maintain a measurable signal intercept. My colleague Dr. Suresh Bhat then overtook the instrument for the final alignment. He enhanced with some minor mechanical modifications the vat stability and the platform repositioning precision. After having reached an intercept of more than 10% on all four detector stations (measured simultaneously), the argon-ion laser (Coherent Innova 300-8) broke down. It was replaced with a HeNe laser which caused a complete realignment of the optical components due to the change of wavelength from $\lambda=514$nm to 632.8nm. When the intercept of all four detector stations (measured simultaenously) reached values of about 5%, this HeNe laser broke down as well - just two weeks before the deadline of this work. Currently, another HeNe laser is used as a coherent light source, and the alignment of the instrument is under way. Unfortunately, none of the previous test measurements was saved, as better results have been expected within a few days and the break-down of two lasers in a row seemed very improbable. Nonetheless, our test measurements demonstrated that the setup is functional. New calibration measurements are expected to be performed within few weeks and published in the coming months.

Chapter 10

Further development/outlook

After having worked intensively on the present Multi 3D set-up and aligning it several times, it became clear that in a next version a number of design problems need to be resolved. Here I try to give a brief summary of the most important modifications that should be considered based on my work. These should of course include a number of additional options in the control and data analysis software (analysis and scripting tools for automated measurements, additional information in the stored data file), but a clear emphasis should be given to some aspects of the hardware where modifications would considerably improve the functionality of the instrument. As the symmetric beamsplitter provides no direct beam, it always has to be removed for lens alignment purposes. It would be much easier to upgrade this "first generation" beamsplitter to the most recent version also used for the commercial version of the 3D instrument that provides a low-intensity direct central beam that is extremely helpful in the alignment procedure. This can be achieved if the prism deflecting the beam downwards in the beamsplitter has a limited transmission of around 10%.

But the biggest draw-back is clearly coming from the dowel pin design of the two platforms. A correct initial alignment of the detector lenses is much more difficult since the split parallel laser beam cannot be brought in the "back position", i.e. the incident laser beams cannot be placed in the position of the detection fibers, illuminating the sample through the detector lens. Therefore the focal length cannot be adjusted. Moreover, a direct beam would considerably help to arrive at an initial guess of a reasonable alignment starting position for tilt and translation. Furthermore, the dowel pins don't ensure sufficiently precise repositioning (the best results were a loss of half of the intercept), and the fact that one platform has to be aligned but then removed again to allow the alignment of the other one makes the alignment very time-consuming. Different approaches seem to be possible. First of all, using lenses with a longer focal length would create space enough to place all detectors on the same side of the direct beam. Therefore it would no longer be necessary to remove an already aligned detector platform because the laser can then be brought in direct line to all detectors. This would also allow for having the detectors mounted on a rotating goniometer arm (or a rotating platform) with a fixed laser mounted in such a way that all detector lenses can be brought into the illuminating beam (once the laser lens is removed). Alternatively the laser could be kept on a moving goniometer arm and the detectors could be mounted individually such that the arm can drive underneath a u-shaped holder. The smallest scattering angle would be aligned first with the bigger ones following sequentially. In both designs, the laser could be brought in the "back position" behind the detector lenses as well as into transmission geometry, facilitating and speeding up the alignment of the instrument considerably.

Bibliography

[1] H. Holthoff, S. U. Egelhaaf, M. Borkovec, P. Schurtenberger, and H. Sticher. Coagulation rate measurements of colloidal particles by simultaneous static and dynamic light scattering. *Langmuir*, 12:5541–5549, 1996.

[2] D. A. Weitz, J. S. Huang, M. Y. Lin, and J. Sung. Limits of the fractal dimension for irreversible kinetic aggregation of gold colloids. *Physical Review Letters*, 54(13):1416, 1985.

[3] D. A. Weitz and M. Oliveira. Fractal structures formed by kinetic aggregation of aequous gold colloids. *Physical Review Letters*, 52(16):1433–1436, April 1984.

[4] B. B. Mandelbrot. *The fractal geometry of nature*. W. H. Freeman and Company, 1982.

[5] A. H. Krall and D. A. Weitz. Internal dynamics and elasticity of fractal colloidal gels. *Physical Review Letters*, 80(4):778–781, 1998.

[6] P. Lindner and Th. Zemb, editors. *Neutrons, X-Rays and Light: Scattering Methods Applied to Soft Condensed Matter*. Elsevier Science, 2002.

[7] H. C. van de Hulst. *Light Scattering by Small Particles*. Dover, 1981.

[8] C. F. Bohren and D. R. Huffman. *Absorption and Scattering of Light by Small Particles*. John Wiley & Sons, 1983.

[9] M. Rottereau, J. C. Gimel, T. Nicolai, and D. Durand. Monte carlos simulation of particle aggregation and gelation: II. pair correlation function and structure factor. *European Physical Journal E*, 15:141–148, 2004.

[10] B. J. Berne and R. Pecora. *Dynamic light scattering: with applications to Chemistry, Biology and Physics*. Wiley, New York, 1976.

[11] P. N. Pusey. *Neutrons, X-Rays and Light: Scattering Methods Applied to Soft Condensed Matter*, chapter 1, pages 3–21. Elsevier Science, 2002.

[12] G. Maret and P. E. Wolf. Effect of brownian motion of scatterers. *Zeitschrift für Physik B*, 65:409–413, 1987.

[13] D. A. Weitz and D. J. Pine. *Diffusing wave spectroscopy*, chapter 16, pages 652–720. Oxford University Press, 1993.

[14] F. Scheffold and P. Schurtenberger. Light scattering probes of viscoelastic fluids and solids. *Soft Materials*, 1(2):139–165, 2003.

[15] K. Schaetzel. Accuracy of photon correlation measurements on nonergodic samples. *Applied Optics*, 32(21):3880–3885, 1993.

[16] P. N. Pusey and W. Van Megen. Dynamic light scattering by non-ergodic media. *Physica A*, 157:705–741, 1989.

[17] S. Romer, F. Scheffold, and P. Schurtenberger. Sol-gel transition of concentrated colloidal suspensions. *Physical Review Letters*, 85(23):4980–4983, 2000.

[18] F. Scheffold, S. E. Skipetrov, S. Romer, and P. Schurtenberger. Diffusing wave spectroscopy of nonergodic media. *Physical Review E*, 63(061404), 2001.

[19] P. Zakharov, F. Cardinaux, and F. Scheffold. Multi-speckle diffusing wave spectroscopy with a single mode detection scheme. *submitted to Physical Review E*, 2005.

[20] K. N. Pham, S. U. Egelhaaf, A. Moussaid, and P. N. Pusey. Ensemble-averaging in dynamic light scattering by an echo technique. *Review of Scientific Instruments*, 75(7):2419–2431, 2004.

[21] A. H. Krall, Z. Huang, and D. A. Weitz. Dynamics of density fluctuations in colloidal gels. *Physica A*, 235:19–33, 1997.

[22] P. Schurtenberger. *Neutrons, X-Rays and Light: Scattering Methods Applied to Soft Condensed Matter*, chapter 7, pages 145–170. Elsevier Science, 2002.

[23] S. Romer, C. Urban, V. Lobaskin, F. Scheffold, A. Stradner, J. Kohlbrecher, and P. Schurtenberger. Simultaneous light and small-angle neutron scattering on aggregating concentrated colloidal suspensions. *Journal of Applied Crystallography*, 36:1–6, 2003.

[24] L. J. Gauckler, Th. Graule, and F. Bader. Ceramic forming using enzyme catalyzed reactions. *Materials Chemistry and Physics*, 2509:1–25, 1999.

[25] J. Kohlbrecher and W. Wagner. The new sans instrument at the swiss spallation source sinq. *Journal of Applied Crystallography*, 33:804–806, 2000.

[26] M. R. Eskildsen, P. L. Gammel, E. D. Isaacs, C. Detlefs, K. Mortensen, and D. J. Bishop. Compound refractive optics for the imaging and focusing of low-energy neutrons. *Letters to Nature*, 391:563–566, 1998.

[27] S. M. Choi, J. G. Barker, C. J. Glinka, Y. T. Cheng, and P. L. Gammel. Focusing cold neutrons with multiple biconcave lenses for small-angle neutron scattering. *Journal of Applied Crystallography*, 33:793–796, 2000.

[28] H. M. Wyss, J. Innerlohinger, L. P. Meier, L. J. Gauckler, and O. Glatter. Small-angle static light scattering of concentrated silica suspensions. *Journal of Colloid and Interface Science*, 271:388–399, 2004.

[29] H. Bissig. *Dynamics of two evolving systems: coarsening foam and attractive colloidal particles*. PhD thesis, University of Fribourg, Switzerland, 2004.

[30] P. N. Segrè, V. Prasad, A. B. Schofield, and D. A. Weitz. Glasslike kinetik arrest at the colloidal-gelation transition. *Physical Review Letters*, 86(26):6042–6045, 2001.

[31] S. Romer, C. Urban, H. Bissig, A. Stradner, F. Scheffold, and P. Schurtenberger. Dynamics of concentrated colloidal suspensions: diffusion, aggregation and gelation. *Philosophical Transactions A*, 35:977–984, 2001.

[32] L. Rojas, S. Romer, F. Scheffold, and P. Schurtenberger. Diffusing wave spectroscopy and small angle neutron scattering from concentrated colloidal suspensions. *Physical Review E*, 65(051403), 2002.

[33] W. Haertl and H. Versmold. An experimental verification of incoherent light scattering. *Journal of Chemical Physics*, 80(4):1387–1389, 1984.

[34] P. N. Pusey. Intensity fluctuation spectroscopy of charged brownian particles: the coherent scattering function. *Journal of Physics A: Mathematical and General*, 11(1):119–135, 1978.

[35] P. N. Pusey, H. M. Fijnaut, and A. Vrij. Mode amplitudes in dynamic light scattering by concentrated liquid suspensions of polydisperes hard spheres. *Journal of Chemical Physics*, 77(9):4270–4281, 1982.

[36] J. C. Brown, P. N. Pusey, J. W. Goodwin, and R. H. Ottewill. Light scattering study of dynamic and time-averaged correlations in dispersions of charged particles. *Journal of Physics A: Mathematical and General*, 8(5):664–682, 1975.

[37] P. Dimon, S. K. Sinha, D. A. Weitz, C. R. Safinya, G. S. Smith, W. A. Varady, and H. M. Lindsay. Structure of aggregated gold colloids. *Physical Review Letters*, 57(5):595–598, 1986.

[38] Jun Liu, Wan Y. Shih, Mehmet Sarikaya, and Ilhan A. Aksay. Fractal colloidal aggregates awith finite interparticle interactions: Energy dependence of the fractal dimension. *Physical Review A*, 41:3206–3213, 1990.

[39] M. Tirado-Miranda, A. Schmitt, J. Callejas-Fernandez, and A. Fernandez-Barbero. Colloidal clusters with finite binding energies: Fractal structure and growth mechanism. *Langmuir*, 15:3437–3444, 1999.

[40] S. Diez Orrite, S. Stoll, and P. Schurtenberger. Off-lattice monte carlo simulations of irreversible and reversible aggregation processes. *submitted to Soft Matter*, 2005.

[41] J. C. Gimel, D. Durand, and T. Nicolai. Transition between flocculation and percolation of a diffusion-limited cluster-cluster aggregation process using three-dimensional monte carlo simulation. *Physical Review B*, 51(17):11348–11357, 1995.

[42] C. J. Brinker and G. W. Scherer. *Sol-Gel Science: The Physics and Chemistry of Sol-Gel Processing*. Academic Press, 1990.

[43] W. Goetze and L. Sjogren. Relaxation processes in supercooled liquids. *Reports on Progress in Physics*, 55:241–376, 1992.

[44] W. van Megen and S. M. Underwood. Glass transition in colloidal hard spheres: Measurements and mode-coupling-theory analysis of the coherent intermediate scattering function. *Physical Review E*, 49(5):4206–4220, 1994.

[45] W. Kob. *Experimental and Theoretical Approaches to Supercooled Liquids: Advances and Novel Applications*, chapter The Mode-Coupling Theory of the Glass Transition, page 12. ACS Books, 1992.

[46] R. G. Larson. *The Structure and Rheology of Complex Fluids*. Oxford University Press, 1999.

[47] K. N. Pham, S. U. Egelhaaf, P. N. Pusey, and W. C. K. Poon. Glasses in hard shperes with short-range attraction. *Physical Review E*, 69:011503, 2004.

[48] S. Romer. *Aggregation and Gelation of Concentrated Colloidal Suspensions*. PhD thesis, Swiss Federal Institute of Technology Zurich, 2001.

[49] P. Poulin, J. Bibette, and D. A. Weitz. From colloidal aggregation to spinodal decomposition in sticky emulsions. *The European Physical Journal B*, 7:277–281, 1999.

[50] C. Urban. *Development of Fiber Optic Based Dynamic Light Scattering for a Characterization of Turbid Suspensions*. PhD thesis, Swiss Federal Institute of Technology Zurich, 1999.

[51] K. Schaetzel. Suppression of multiple scattering by photon cross-correlation techniques. *Journal of Modern Optics*, 38(9):1849–1865, 1991.

[52] R. Finsy, P. de Groen, L. Deriemaeker, E. Gelade, and J. Joosten. Data analysis of multi-angle photon correlation measurements without and with prior knowledge. *Particle and Particle Systems Characterization*, 9(1-4):237–251, 1992.

[53] F. Scheffold, A. Shalkevich, R. Vavrin, J. Crassous, and P. Schurtenberger. *Particle Sizing and Characterization*, chapter PCS particle sizing in turbid suspensions: scope and limitations, pages 1–28. Number 881 in ACS Symposium Series. Oxford University Press, 2004.

[54] P. N. Pusey. Suppression of multiple scattering by photon cross-correlation techniques. *Current Opinion in Colloid and Interface Science*, 4:177–185, 1999.

[55] K. Schaetzel, M. Drewel, and J. Ahrens. Suppression of multiple scattering in photon correlation spectroscopy. *Journal of Physics: Condensed Matter*, 2:SA393–SA398, 1990.

[56] E. Overbeck, C. Sinn, T. Palberg, and K. Schaetzel. Probing dynamics of dense suspensions: three-dimensional cross-correlation technique. *Colloids and Surfaces A: Physicochemical and Engineering Aspects*, 122(1-3):83–87, 1997.

[57] L. B. Aberle, P. Huelstede, S. Wiegand, W. Schroer, and W. Staude. Effective suppression of multiply scattered light in static and dynamic light scattering. *Applied Optics*, 37(27):6511–6524, 1998.

[58] C. Urban and P. Schurtenberger. Characterization of turbid colloidal suspensions using light scattering techniques combined with cross-correlation methods. *Journal of Colloid and Interface Science*, 207:150–158, 1998.

[59] E. Overbeck and C. Sinn. Three-dimensional dynamic light scattering. *Journal of Modern Optics*, 46(2):303–326, 1999.

Acknowledgments

I am grateful to Prof. P. Schurtenberger who gave me the opportunity to conduct this work and who provided guidance and shared his expertise throughout the writing of this thesis.

Many thanks to Prof. F. Scheffold and V. Trappe who supervised this work for fruitful and substantial discussions and their energy and time.

I thank Prof. J. F. Van der Veen and Prof. F. Scheffold for agreeing to be co-examiners of this work.

Thanks a lot to C. Urban for his help and for providing an essential knowledge transfer not only concerning the 3D technique.

Many thanks to A. Stradner for her help with the SANS measurements and many thanks to the local contact at PSI, J. Kohlbrecher.

Thanks a lot to the MM group in Fribourg, especially S. Bhat for his help with the Multi3D, F. Mueller for the help with measurements and for numerous discussions, F. Cardinaux and P. Zakharov for their help with the Echo measurements, L. Rojas for the l^* calculations, J. Clara Rahola for the DLVO potential calculations and H. Bissig for his help with Urea/Urease process, DWS and for being the ideal office mate.

Many thanks as well to the Mechanical and Electrical Workshop at the Physics Faculty of the University of Fribourg for their competent help.

Die VDM Verlagsservicegesellschaft sucht für wissenschaftliche Verlage abgeschlossene und herausragende

Dissertationen, Habilitationen, Diplomarbeiten, Master Theses, Magisterarbeiten usw.

für die kostenlose Publikation als Fachbuch.

Sie verfügen über eine Arbeit, die hohen inhaltlichen und formalen Ansprüchen genügt, und haben Interesse an einer honorarvergüteten Publikation?

Dann senden Sie bitte erste Informationen über sich und Ihre Arbeit per Email an *info@vdm-vsg.de*.

Sie erhalten kurzfristig unser Feedback!

VDM Verlagsservicegesellschaft mbH
Dudweiler Landstr. 99　　　　　　　Telefon +49 681 3720 174
D - 66123 Saarbrücken　　　　　　　Fax　　　+49 681 3720 1749
www.vdm-vsg.de

Die VDM Verlagsservicegesellschaft mbH vertritt

Printed by Books on Demand GmbH, Norderstedt / Germany